# Essential Facts and Tables

Published with the permission of R.I.C. Publications Pty. Ltd.



First published by R.I.C. Publications Pty. Ltd., Perth, Western Australia. Revised by Didax Educational Resources.



Printed in the United States of America.

Order Number 2-5178
ISBN 1-58324-112-4

B C D E F 09 08 07 06 05

Didax Educational Resources

395 Main Street
Rowley, MA 01969
www.worldteacherspress.com

## CONTENTS

## SYMBOLS

| Symbol | Meaning | Symbol | Meaning |
|---|---|---|---|
| + | addition | △ | triangle |
| − | subtraction | ∪ | union |
| x | multiplication | ∩ | intersection |
| ÷ | division | π | pi (3.14) |
| > | greater than | {([ ])} | brackets |
| < | less than | ∞ | infinity |
| = | equal to | # | number |
| ≠ | not equal to | ∴ | therefore |
| ≈ | approximately equal to | ∵ | because |
| ° | degree | % | percent |
| ∠ | angle | √ | square root |
| ∟ | right angle | . | decimal point |
| ⊥ | is perpendicular to | : | ratio |
| ═ | parallel | ¢ | cent |
| ≅ | is congruent to | $ | dollar |

| Length | |
|---|---|
| mm | millimeter |
| cm | centimeter |
| m | meter |
| km | kilometer |

| Area | |
|---|---|
| $mm^2$ | millimeter squared |
| $cm^2$ | centimeter squared |
| $m^2$ | meter squared |
| $km^2$ | kilometer squared |
| ha | hectare |

| Capacity | |
|---|---|
| mL | milliliter |
| L | liter |
| kL | kiloliter |

| Mass | |
|---|---|
| g | gram |
| kg | kilogram |
| t | ton |

| Temperature | |
|---|---|
| °C | degrees Celsius |
| °F | degrees Fahrenheit |
| °K | degrees Kelvin |

## *MATH RULES AND FORMULAS*

**Circle**

**Circumference** = 2R

**Area** = $R^2$

**Rectangle**

**Perimeter** = 2 x (L + W)

**Area** = L x W

**Triangle**

**Area** = $\frac{1}{2}$ x (B x H)

**Square**

**Perimeter** = 4 x L

**Area** = $L^2$

R = radius of circle
L = length of shape
W = width of shape
B = base (length)
H = perpendicular height of triangle

## *PARTS OF A CALCULATOR*

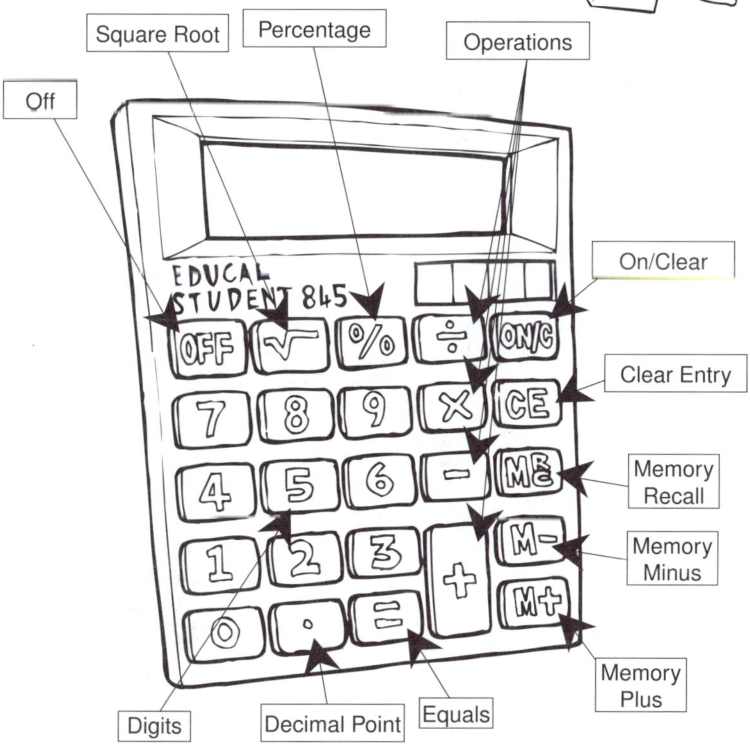

## ROMAN NUMERALS

| I | V | X | L | C | D | M |
|---|---|---|---|---|---|---|
| 1 | 5 | 10 | 50 | 100 | 500 | 1,000 |

| | |
|---|---|
| 1 | **I** |
| 2 | **II** |
| 3 | **III** |
| 4 | **IV** |
| 5 | **V** |
| 6 | **VI** |
| 7 | **VII** |
| 8 | **VIII** |
| 9 | **IX** |
| 10 | **X** |

| | |
|---|---|
| 11 | **XI** |
| 12 | **XII** |
| 13 | **XIII** |
| 14 | **XIV** |
| 15 | **XV** |
| 16 | **XVI** |
| 17 | **XVII** |
| 18 | **XVIII** |
| 19 | **XIX** |
| 20 | **XX** |

| | |
|---|---|
| 21 | **XXI** |
| 22 | **XXII** |
| 23 | **XXIII** |
| 24 | **XXIV** |
| 25 | **XXV** |
| 26 | **XXVI** |
| 27 | **XXVII** |
| 28 | **XXVIII** |
| 29 | **XXIX** |
| 30 | **XXX** |

| | |
|---|---|
| 35 | **XXXV** |
| 40 | **XL** |
| 45 | **XLV** |
| 50 | **L** |
| 55 | **LV** |
| 60 | **LX** |
| 65 | **LXV** |
| 70 | **LXX** |
| 75 | **LXXV** |
| 80 | **LXXX** |
| 85 | **LXXXV** |
| 90 | **XC** |
| 95 | **XCV** |

| | |
|---|---|
| 100 | **C** |
| 150 | **CL** |
| 200 | **CC** |
| 300 | **CCC** |
| 400 | **CD** |
| 500 | **D** |
| 600 | **DC** |
| 700 | **DCC** |
| 800 | **DCCC** |
| 900 | **CM** |
| 1,000 | **M** |
| 1,500 | **MD** |
| 2,000 | **MM** |

## INFINITY

Infinity is a word derived from the word *infinite* which means...
**...having no limit, endless, impossible to measure.**

The symbol used for infinity is $\infty$.

It is a word used to describe a situation that has no end, such as:

- the number of ways you can paint a painting;
- counting from 1, 2, 3, 4, . . . to infinity; or
- traveling into space.

## PRIME NUMBERS

A prime number is...
**...a number that can be divided evenly by only 1 and itself.**

Examples of prime numbers are:

2, 3, 5, 7, 11, 13 and 17.

1 is not a prime number. It is a special number because it doesn't fit into any particular category.

## COMPOSITE NUMBERS

A composite number is...
**...a number that can be divided by more than itself and 1.**

Examples of composite numbers are:

4, 6, 8, 9, 10, 12 and 14.

## PATTERNS

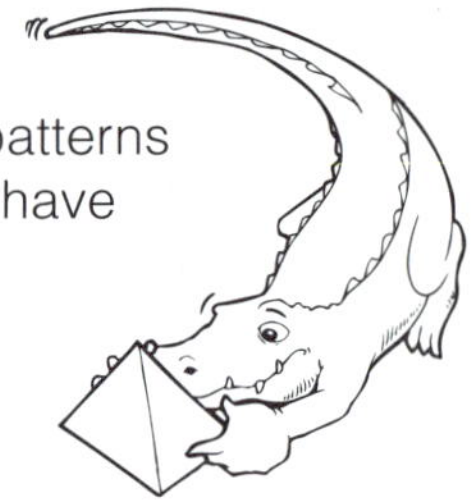

Patterns occur in numbers in many different ways. Mathematicians have been studying these number patterns for centuries and the discoveries from ancient times have helped us to form our modern view of mathematics.

Simple number patterns such as odd and even numbers are easily identified; however, there are many more observable complex patterns. Some of these are listed below.

### TRIANGULAR NUMBERS

**...numbers that can form the shape of a triangle**

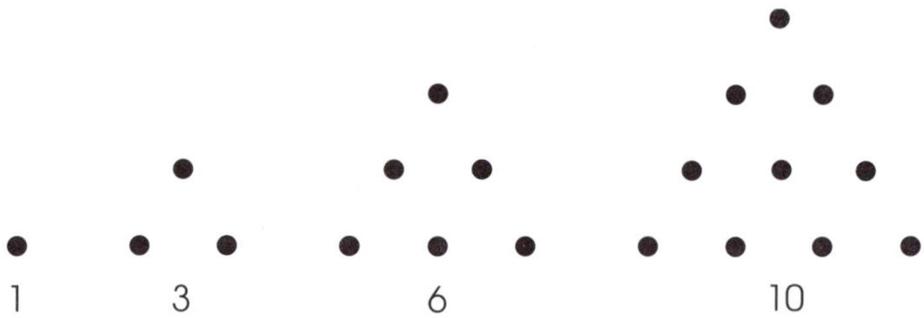


### SQUARE NUMBERS

**...numbers that can form the shape of a square**

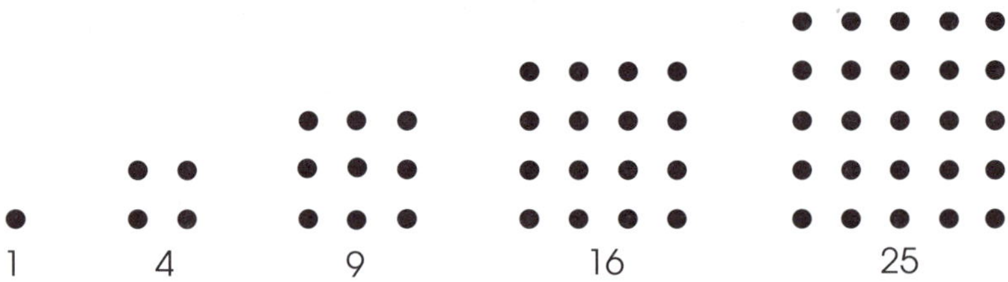


Patterns can be identified in many other ways. Some of these include:

- rectangular numbers;
- hexagonal numbers;
- pentagonal numbers; and
- cubic numbers.

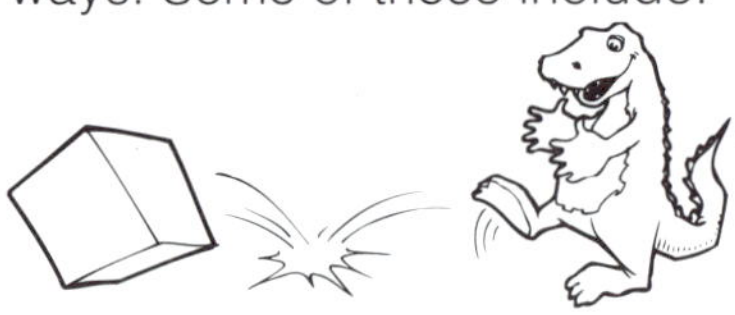

Experiment to find what patterns exist in these and other numbers.

## *INDEX NOTATION, FACTORS AND MULTIPLES*

### *INDEX NOTATION*

**...power or index notation is a short way of writing large numbers by multiplication of repeated factors**

The power of any number is also called **index** or **exponential** notation.

*For example:*

$8^2$ is the same as 8 x 8

$8^3$ is the same as 8 x 8 x 8

$8^5$ is the same as 8 x 8 x 8 x 8 x 8

$8^2$ — exponent (2), base (8)

Index notation is also called **power** as this is the word used to describe a number.

*For example:*

$8^2$ is called 8 to the power of two (*also called 8 squared*)

$8^3$ is called 8 to the power of three (*also called 8 cubed*)

$8^5$ is called 8 to the power of five

and so on…

### *FACTORS*

**...a factor of a number is a number that will divide evenly into that number**

All numbers except 1 have more than one factor.

*For example:*
12 is the number
The factors of 12 are 1, 2, 3, 4, 6 and 12

**...a prime factor is a prime number that will divide evenly into a given number**

*For example:*
2, 3 and 5 are prime factors of 30

### *MULTIPLES*

**...a multiple of a number is that number multiplied by other whole numbers**

*For example:*
the multiples of 5 are: 5, 10, 15, 20, 25, 30, 35, 40, 45,…

## ADDITION AND SUBTRACTION FACTS

To help understand addition and subtraction facts, always remember, just like multiplication and division, if you know one fact then you will know several others.

*For example:*
If you know 4 + 5 = 9
then you also know
5 + 4 = 9 and 9 – 5 = 4 and 9 – 4 = 5

This table will help you with all your addition and subtraction facts to 20.

| + | 0 | 1 | 2 | 3 | 4 | 5 | 6 | 7 | 8 | 9 | 10 |
|---|---|---|---|---|---|---|---|---|---|---|---|
| **0** | 0 | 1 | 2 | 3 | 4 | 5 | 6 | 7 | 8 | 9 | 10 |
| **1** | 1 | 2 | 3 | 4 | 5 | 6 | 7 | 8 | 9 | 10 | 11 |
| **2** | 2 | 3 | 4 | 5 | 6 | 7 | 8 | 9 | 10 | 11 | 12 |
| **3** | 3 | 4 | 5 | 6 | 7 | 8 | 9 | 10 | 11 | 12 | 13 |
| **4** | 4 | 5 | 6 | 7 | 8 | 9 | 10 | 11 | 12 | 13 | 14 |
| **5** | 5 | 6 | 7 | 8 | 9 | 10 | 11 | 12 | 13 | 14 | 15 |
| **6** | 6 | 7 | 8 | 9 | 10 | 11 | 12 | 13 | 14 | 15 | 16 |
| **7** | 7 | 8 | 9 | 10 | 11 | 12 | 13 | 14 | 15 | 16 | 17 |
| **8** | 8 | 9 | 10 | 11 | 12 | 13 | 14 | 15 | 16 | 17 | 18 |
| **9** | 9 | 10 | 11 | 12 | 13 | 14 | 15 | 16 | 17 | 18 | 19 |
| **10** | 10 | 11 | 12 | 13 | 14 | 15 | 16 | 17 | 18 | 19 | 20 |

## FRACTIONS, DECIMALS AND PERCENTAGES

### FRACTIONS

**A fraction is a number which usually represents a part of a whole number.**

It is made up of:

- a **denominator**, which is the number below the line indicating how many parts the whole number is divided into.
- a **numerator**, which is the number above the line indicating how many parts are in consideration.

$$\frac{3}{4}$$

numerator (3)

denominator (4)

- a **proper** fraction is a fraction where the value of the numerator is smaller than the denominator. *For example:* $\frac{1}{2}$
- An **improper fraction** is a fraction where the numerator is larger than the denominator. *For example:* $\frac{4}{3}$
- A **mixed number** is both a whole number and a proper fraction. *For example:* $1\frac{2}{3}$

### DECIMALS

**The decimal number system is based on the number ten.**

The decimal place system is based on multiples of ten where a whole number is divided into tenths, hundredths, thousandths and so on.

This diagram shows how we use decimal fractions, or the decimal place system.

1 whole number

↓

divided into ten equal parts

| 0.1 | 0.1 | 0.1 | 0.1 | 0.1 | 0.1 | 0.1 | 0.1 | 0.1 | 0.1 |
|---|---|---|---|---|---|---|---|---|---|

each tenth can be divided into a further ten equal parts to make hundredths

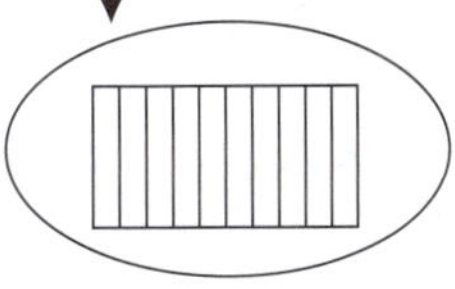

## *FRACTIONS, DECIMALS AND PERCENTAGES*

### PERCENTAGES

**A percentage is a number or quantity represented in hundredths.**

*For example:*

If you achieved 8 correct answers out of a possible 10 in a spelling test, then your percentage is 80%.

To convert a number or fraction to a percentage it is necessary to multiply the number by 100.

*For example:*

$$\frac{8}{10} \text{ X } \frac{100}{1} = \frac{800}{10} = 80\%$$

### CONVERSION CHART

| Fraction | Decimal | Percentage |
|---|---|---|
| $\frac{1}{2}$ | 0.5 | 50% |
| $\frac{1}{3}$ | $0.3\bar{3}$ | 33.33% |
| $\frac{1}{4}$ | 0.25 | 25% |
| $\frac{1}{5}$ | 0.2 | 20% |
| $\frac{1}{8}$ | 0.125 | 12.5% |
| $\frac{1}{10}$ | 0.1 | 10% |
| $\frac{1}{20}$ | 0.05 | 5% |
| $\frac{1}{25}$ | 0.04 | 4% |
| $\frac{1}{50}$ | 0.02 | 2% |

## MULTIPLICATION AND DIVISION FACTS

Multiplication and division facts are linked together. If you know one multiplication fact you also know three other related facts.

*For example:*

5 x 3 = 15

If this basic fact is known, you also know:

3 x 5 = 15 and 15 ÷ 3 = 5 and 15 ÷ 5 = 3

The following tables include the times tables and other facts you will learn at the same time.

| 2 x table |
| --- |
| **1 x 2 = 2** |
| **2 x 2 = 4** |
| **3 x 2 = 6** |
| **4 x 2 = 8** |
| **5 x 2 = 10** |
| **6 x 2 = 12** |
| **7 x 2 = 14** |
| **8 x 2 = 16** |
| **9 x 2 = 18** |
| **10 x 2 = 20** |

| other facts I know | | |
| --- | --- | --- |
| 2 x 1 = 2 | 2 ÷ 2 = 1 | 2 ÷ 1 = 2 |
| | 4 ÷ 2 = 2 | |
| 2 x 3 = 6 | 6 ÷ 2 = 3 | 6 ÷ 3 = 2 |
| 2 x 4 = 8 | 8 ÷ 2 = 4 | 8 ÷ 4 = 2 |
| 2 x 5 = 10 | 10 ÷ 2 = 5 | 10 ÷ 5 = 2 |
| 2 x 6 = 12 | 12 ÷ 2 = 6 | 12 ÷ 6 = 2 |
| 2 x 7 = 14 | 14 ÷ 2 = 7 | 14 ÷ 7 = 2 |
| 2 x 8 = 16 | 16 ÷ 2 = 8 | 16 ÷ 8 = 2 |
| 2 x 9 = 18 | 18 ÷ 2 = 9 | 18 ÷ 9 = 2 |
| 2 x 10 = 20 | 20 ÷ 2 = 10 | 20 ÷ 10 = 2 |

## MULTIPLICATION AND DIVISION FACTS

| 3 x table |
|---|
| 1 x 3 = 3 |
| 2 x 3 = 6 |
| 3 x 3 = 9 |
| 4 x 3 = 12 |
| 5 x 3 = 15 |
| 6 x 3 = 18 |
| 7 x 3 = 21 |
| 8 x 3 = 24 |
| 9 x 3 = 27 |
| 10 x 3 = 30 |

| other facts I know | | |
|---|---|---|
| 3 x 1 = 3 | 3 ÷ 3 = 1 | 3 ÷ 1 = 3 |
| 3 x 2 = 6 | 6 ÷ 3 = 2 | 6 ÷ 2 = 3 |
| | 9 ÷ 3 = 3 | |
| 3 x 4 = 12 | 12 ÷ 3 = 4 | 12 ÷ 4 = 3 |
| 3 x 5 = 15 | 15 ÷ 3 = 5 | 15 ÷ 5 = 3 |
| 3 x 6 = 18 | 18 ÷ 3 = 6 | 18 ÷ 6 = 3 |
| 3 x 7 = 21 | 21 ÷ 3 = 7 | 21 ÷ 7 = 3 |
| 3 x 8 = 24 | 24 ÷ 3 = 8 | 24 ÷ 8 = 3 |
| 3 x 9 = 27 | 27 ÷ 3 = 9 | 27 ÷ 9 = 3 |
| 3 x 10 = 30 | 30 ÷ 3 = 10 | 30 ÷ 10 = 3 |

| 4 x table |
|---|
| 1 x 4 = 4 |
| 2 x 4 = 8 |
| 3 x 4 = 12 |
| 4 x 4 = 16 |
| 5 x 4 = 20 |
| 6 x 4 = 24 |
| 7 x 4 = 28 |
| 8 x 4 = 32 |
| 9 x 4 = 36 |
| 10 x 4 = 40 |

| other facts I know | | |
|---|---|---|
| 4 x 1 = 4 | 4 ÷ 4 = 1 | 4 ÷ 1 = 4 |
| 4 x 2 = 8 | 8 ÷ 4 = 2 | 8 ÷ 2 = 4 |
| 4 x 3 = 12 | 12 ÷ 4 = 3 | 12 ÷ 3 = 4 |
| | 16 ÷ 4 = 4 | |
| 4 x 5 = 20 | 20 ÷ 4 = 5 | 20 ÷ 5 = 4 |
| 4 x 6 = 24 | 24 ÷ 4 = 6 | 24 ÷ 6 = 4 |
| 4 x 7 = 28 | 28 ÷ 4 = 7 | 28 ÷ 7 = 4 |
| 4 x 8 = 32 | 32 ÷ 4 = 8 | 32 ÷ 8 = 4 |
| 4 x 9 = 36 | 36 ÷ 4 = 9 | 36 ÷ 9 = 4 |
| 4 x 10 = 40 | 40 ÷ 4 = 10 | 40 ÷ 10 = 4 |

## MULTIPLICATION AND DIVISION FACTS

| 5 x table | other facts I know | | |
|---|---|---|---|
| 1 x 5 = 5 | 5 x 1 = 5 | 5 ÷ 5 = 1 | 5 ÷ 1 = 5 |
| 2 x 5 = 10 | 5 x 2 = 10 | 10 ÷ 5 = 2 | 10 ÷ 2 = 5 |
| 3 x 5 = 15 | 5 x 3 = 15 | 15 ÷ 5 = 3 | 15 ÷ 3 = 5 |
| 4 x 5 = 20 | 5 x 4 = 20 | 20 ÷ 5 = 4 | 20 ÷ 4 = 5 |
| 5 x 5 = 25 | | 25 ÷ 5 = 5 | |
| 6 x 5 = 30 | 5 x 6 = 30 | 30 ÷ 5 = 6 | 30 ÷ 6 = 5 |
| 7 x 5 = 35 | 5 x 7 = 35 | 35 ÷ 5 = 7 | 35 ÷ 7 = 5 |
| 8 x 5 = 40 | 5 x 8 = 40 | 40 ÷ 5 = 8 | 40 ÷ 8 = 5 |
| 9 x 5 = 45 | 5 x 9 = 45 | 45 ÷ 5 = 9 | 45 ÷ 9 = 5 |
| 10 x 5 = 50 | 5 x 10 = 50 | 50 ÷ 5 = 10 | 50 ÷ 10 = 5 |

| 6 x table | other facts I know | | |
|---|---|---|---|
| 1 x 6 = 6 | 6 x 1 = 6 | 6 ÷ 6 = 1 | 6 ÷ 1 = 6 |
| 2 x 6 = 12 | 6 x 2 = 12 | 12 ÷ 6 = 2 | 12 ÷ 2 = 6 |
| 3 x 6 = 18 | 6 x 3 = 18 | 18 ÷ 6 = 3 | 18 ÷ 3 = 6 |
| 4 x 6 = 24 | 6 x 4 = 24 | 24 ÷ 6 = 4 | 24 ÷ 4 = 6 |
| 5 x 6 = 30 | 6 x 5 = 30 | 30 ÷ 6 = 5 | 30 ÷ 5 = 6 |
| 6 x 6 = 36 | | 36 ÷ 6 = 6 | |
| 7 x 6 = 42 | 6 x 7 = 42 | 42 ÷ 6 = 7 | 42 ÷ 7 = 6 |
| 8 x 6 = 48 | 6 x 8 = 48 | 48 ÷ 6 = 8 | 48 ÷ 8 = 6 |
| 9 x 6 = 54 | 6 x 9 = 54 | 54 ÷ 6 = 9 | 54 ÷ 9 = 6 |
| 10 x 6 = 60 | 6 x 10 = 60 | 60 ÷ 6 = 10 | 60 ÷ 10 = 6 |

## MULTIPLICATION AND DIVISION FACTS

## 7

| 7 x table |
|---|
| **1 x 7 = 7** |
| **2 x 7 = 14** |
| **3 x 7 = 21** |
| **4 x 7 = 28** |
| **5 x 7 = 35** |
| **6 x 7 = 42** |
| **7 x 7 = 49** |
| **8 x 7 = 56** |
| **9 x 7 = 63** |
| **10 x 7 = 70** |

| other facts I know | | |
|---|---|---|
| 7 x 1 = 7 | 7 ÷ 7 = 1 | 7 ÷ 1 = 7 |
| 7 x 2 = 14 | 14 ÷ 7 = 2 | 14 ÷ 2 = 7 |
| 7 x 3 = 21 | 21 ÷ 7 = 3 | 21 ÷ 3 = 7 |
| 7 x 4 = 28 | 28 ÷ 7 = 4 | 28 ÷ 4 = 7 |
| 7 x 5 = 35 | 35 ÷ 7 = 5 | 35 ÷ 5 = 7 |
| 7 x 6 = 42 | 42 ÷ 7 = 6 | 42 ÷ 6 = 7 |
| | 49 ÷ 7 = 7 | |
| 7 x 8 = 56 | 56 ÷ 7 = 8 | 56 ÷ 8 = 7 |
| 7 x 9 = 63 | 63 ÷ 7 = 9 | 63 ÷ 9 = 7 |
| 7 x 10 = 70 | 70 ÷ 7 = 10 | 70 ÷ 10 = 7 |

## 8

| 8 x table |
|---|
| **1 x 8 = 8** |
| **2 x 8 = 16** |
| **3 x 8 = 24** |
| **4 x 8 = 32** |
| **5 x 8 = 40** |
| **6 x 8 = 48** |
| **7 x 8 = 56** |
| **8 x 8 = 64** |
| **9 x 8 = 72** |
| **10 x 8 = 80** |

| other facts I know | | |
|---|---|---|
| 8 x 1 = 8 | 8 ÷ 8 = 1 | 8 ÷ 1 = 8 |
| 8 x 2 = 16 | 16 ÷ 8 = 2 | 16 ÷ 2 = 8 |
| 8 x 3 = 24 | 24 ÷ 8 = 3 | 24 ÷ 3 = 8 |
| 8 x 4 = 32 | 32 ÷ 8 = 4 | 32 ÷ 4 = 8 |
| 8 x 5 = 40 | 40 ÷ 8 = 5 | 40 ÷ 5 = 8 |
| 8 x 6 = 48 | 48 ÷ 8 = 6 | 48 ÷ 6 = 8 |
| 8 x 7 = 56 | 56 ÷ 8 = 7 | 56 ÷ 7 = 8 |
| | 64 ÷ 8 = 8 | |
| 8 x 9 = 72 | 72 ÷ 8 = 9 | 72 ÷ 9 = 8 |
| 8 x 10 = 80 | 80 ÷ 8 = 10 | 80 ÷ 10 = 8 |

# 9 MULTIPLICATION AND DIVISION FACTS

| 9 x table | other facts I know | | |
|---|---|---|---|
| 1 x 9 = 9 | 9 x 1 = 9 | 9 ÷ 9 = 1 | 9 ÷ 1 = 9 |
| 2 x 9 = 18 | 9 x 2 = 18 | 18 ÷ 9 = 2 | 18 ÷ 2 = 9 |
| 3 x 9 = 27 | 9 x 3 = 27 | 27 ÷ 9 = 3 | 27 ÷ 3 = 9 |
| 4 x 9 = 36 | 9 x 4 = 36 | 36 ÷ 9 = 4 | 36 ÷ 4 = 9 |
| 5 x 9 = 45 | 9 x 5 = 45 | 45 ÷ 9 = 5 | 45 ÷ 5 = 9 |
| 6 x 9 = 54 | 9 x 6 = 54 | 54 ÷ 9 = 6 | 54 ÷ 6 = 9 |
| 7 x 9 = 63 | 9 x 7 = 63 | 63 ÷ 9 = 7 | 63 ÷ 7 = 9 |
| 8 x 9 = 72 | 9 x 8 = 72 | 72 ÷ 9 = 8 | 72 ÷ 8 = 9 |
| 9 x 9 = 81 | | 81 ÷ 9 = 9 | |
| 10 x 9 = 90 | 9 x 10 = 90 | 90 ÷ 9 = 10 | 90 ÷ 10 = 9 |

## 10

| 10 x table | other facts I know | | |
|---|---|---|---|
| 1 x 10 = 10 | 10 x 1 = 10 | 10 ÷ 10 = 1 | 10 ÷ 1 = 10 |
| 2 x 10 = 20 | 10 x 2 = 20 | 20 ÷ 10 = 2 | 20 ÷ 2 = 10 |
| 3 x 10 = 30 | 10 x 3 = 30 | 30 ÷ 10 = 3 | 30 ÷ 3 = 10 |
| 4 x 10 = 40 | 10 x 4 = 40 | 40 ÷ 10 = 4 | 40 ÷ 4 = 10 |
| 5 x 10 = 50 | 10 x 5 = 50 | 50 ÷ 10 = 5 | 50 ÷ 5 = 10 |
| 6 x 10 = 60 | 10 x 6 = 60 | 60 ÷ 10 = 6 | 60 ÷ 6 = 10 |
| 7 x 10 = 70 | 10 x 7 = 70 | 70 ÷ 10 = 7 | 70 ÷ 7 = 10 |
| 8 x 10 = 80 | 10 x 8 = 80 | 80 ÷ 10 = 8 | 80 ÷ 8 = 10 |
| 9 x 10 = 90 | 10 x 9 = 90 | 90 ÷ 10 = 9 | 90 ÷ 9 = 10 |
| 10 x 10 = 100 | | 100 ÷ 10 = 10 | |

## THE ESSENTIAL FACTS

Basic facts repeat themselves throughout each of the number groupings.

*For example:*

2 x 3 = 6 is found in the two times table.

It is also found in the three times table.

This makes one less table or basic fact you need to learn.

This happens more and more as you move up the tables and, because of this, assuming the one times table and ten times table are excluded, there are only 36 actual multiplication facts you need to learn in order to know all of your tables. These are listed below.

2 x 2 = 4
3 x 2 = 6
4 x 2 = 8
5 x 2 = 10
6 x 2 = 12
7 x 2 = 14
8 x 2 = 16
9 x 2 = 18

3 x 3 = 9
4 x 3 = 12
5 x 3 = 15
6 x 3 = 18
7 x 3 = 21
8 x 3 = 24
9 x 3 = 27

4 x 4 = 16
5 x 4 = 20
6 x 4 = 24
7 x 4 = 28
8 x 4 = 32
9 x 4 = 36

5 x 5 = 25
6 x 5 = 30
7 x 5 = 35
8 x 5 = 40
9 x 5 = 45

6 x 6 = 36
7 x 6 = 42
8 x 6 = 48
9 x 6 = 54

7 x 7 = 49
8 x 7 = 56
9 x 7 = 63

8 x 8 = 64
9 x 8 = 72

9 x 9 = 81

## MEASUREMENT

The two main systems of measurement used throughout the world are the **metric** and the **imperial**. In the majority of other countries, the metric system is used.

The metric system is a decimal system of measure.

### LENGTH

The following are the units used to measure length.

| unit | abbreviation |
|---|---|
| millimeter | mm |
| centimeter | cm |
| meter | m |
| kilometer | km |

| | | | |
|---|---|---|---|
| 10 | millimeters | = | 1 centimeter |
| 100 | centimeters | = | 1 meter |
| 1,000 | meters | = | 1 kilometer |

Measurements should always be written in only one unit of measure.

*For example:*

Correct: 2.75 m

Incorrect: 2 m, 7 cm, 5 mm

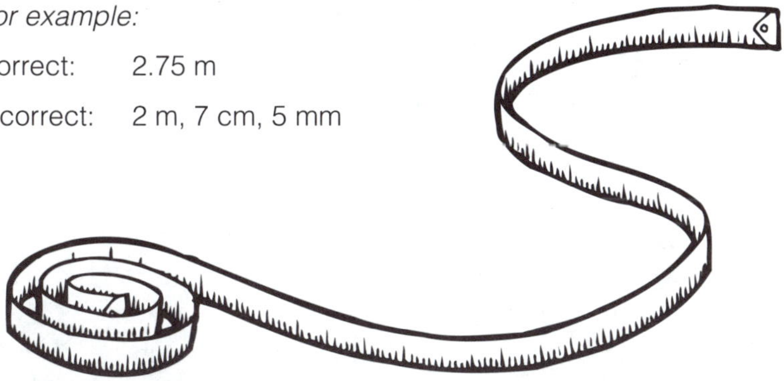

## MEASUREMENT

### AREA

The following are the units used to measure area.

| unit | abbreviation |
|---|---|
| square millimeter | $mm^2$ |
| square centimeter | $cm^2$ |
| square meter | $m^2$ |
| hectare | ha |
| square kilometer | $km^2$ |

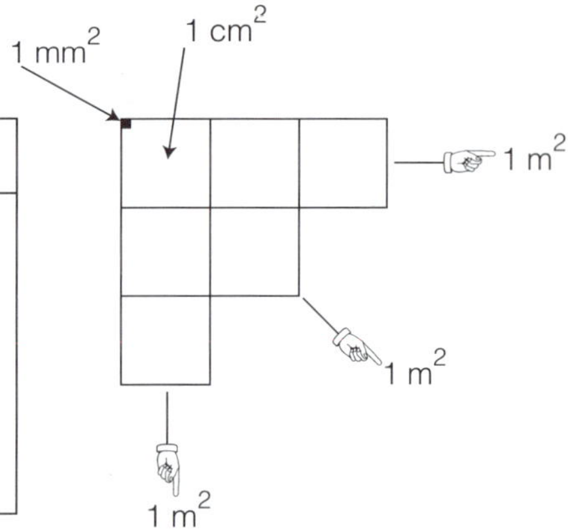


100 millimeters squared = 1 centimeter squared

10,000 centimeters squared = 1 meter squared

10,000 meters squared = 1 hectare

100 hectares = 1 kilometer squared

(These can be spoken "square centimeter," "square meter" or "square kilometer.")

### MASS

The following are the units used to measure mass.

| unit | abbreviation |
|---|---|
| gram | g |
| kilogram | kg |
| ton | t |

1,000 grams = 1 kilogram

1,000 kilograms = 1 ton

## VOLUME

The amount of space taken up by an object.

The following are the units used to measure volume.

| unit | abbreviation |
|---|---|
| cubic millimeter | $mm^3$ |
| cubic centimeter | $cm^3$ |
| cubic meter | $m^3$ |

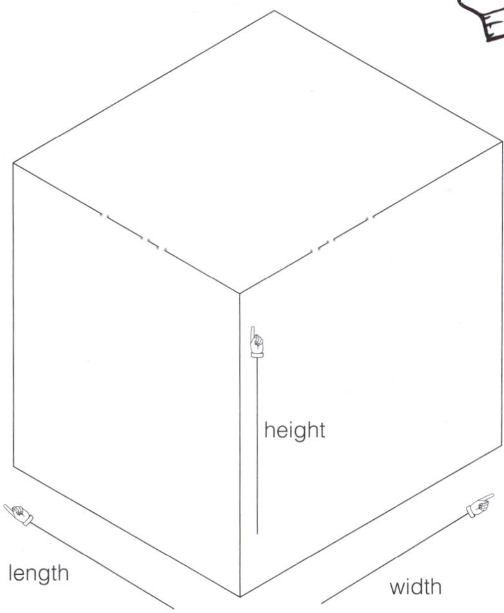


The three dimensions of a cubic measure are:

length height width

| | | | |
|---|---|---|---|
| 1,000 | cubic millimeters | = | 1 cubic centimeter |
| 1,000,000 | cubic centimeters | = | 1 cubic meter |

*MEASUREMENT*

## CAPACITY

The amount of space in the interior of an object measured in cubic units.

The following are the units used to measure capacity.

| unit | abbreviation |
|---|---|
| milliliter | mL |
| liter | L |
| kiloliter | kL |

1,000 milliliters = 1 liter

1,000 liters = 1 kiloliter

## TEMPERATURE

Temperature is measured in either Celsius or Fahrenheit. In the United States, Farenheit is the measure used.

Farenheit measure is recorded in this way:

24°F

Two significant measures in Farenheit measurement are the freezing and boiling points of water.

Freezing Point: 32°F

Boiling Point: 212°F

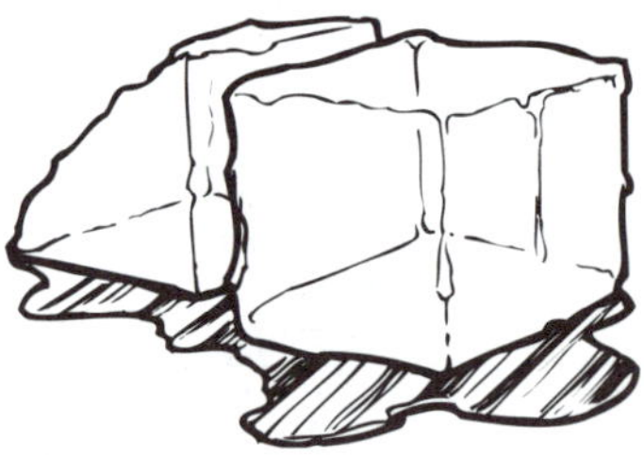

## TIME

The measurement of time is not a decimal system of measure.

| unit | abbreviation |
|---|---|
| second | sec. |
| minute | min. |
| hour | hr. or h. |
| day | d. |
| month | mo. |
| year | yr. |
| century | C |
| millennium | M |

| | | |
|---|---|---|
| 60 seconds | = | 1 minute |
| 60 minutes | = | 1 hour |
| 24 hours | = | 1 day |
| 7 days | = | 1 week |
| 52 weeks | = | 1 year |
| 28–31 days | = | 1 month |
| 365 days | = | 1 year |
| 366 days | = | 1 leap year |
| 10 years | = | 1 decade |
| 100 years | = | 1 century |
| 1,000 years | = | 1 millennium |

*MEASUREMENT*

## TIME

| | |
|---|---|
| January | 31 days |
| February | 28 days (29 in a leap year) |
| March | 31 days |
| April | 30 days |
| May | 31 days |
| June | 30 days |
| July | 31 days |
| August | 31 days |
| September | 30 days |
| October | 31 days |
| November | 30 days |
| December | 31 days |

| | Southern Hemisphere | Northern Hemisphere |
|---|---|---|
| **Summer** | December<br>January<br>February | June<br>July<br>August |
| **Autumn** | March<br>April<br>May | September<br>October<br>November |
| **Winter** | June<br>July<br>August | December<br>January<br>February |
| **Spring** | September<br>October<br>November | March<br>April<br>May |

## MONEY – COINS

10¢

50¢

5¢

1¢

25¢

## MONEY – BILLS

**$1.00**

**$5.00**

**$10.00**

**$20.00**

**$100.00**

## *LINES*

### Line

Made up of an infinite set of points extending in both directions.

### Ray

Made up of an infinite set of points emanating from a point and going in one direction.

### Line Segment

A line with two end points.

### Diagonal

A straight line joining two non-adjacent vertices of a polygon.

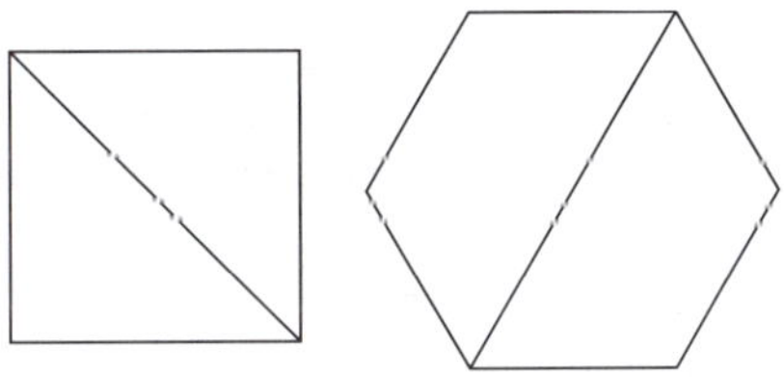

### Horizontal

A line parallel to the horizon.

### Vertical

A line which is at right angles to a horizontal line.

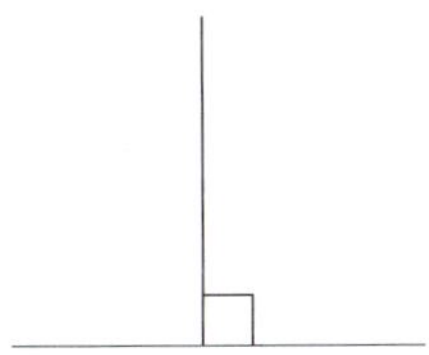

### Perpendicular

Lines at right angles to another line or plane.

### Parallel

Lines that are always the same distance apart and have no common points.

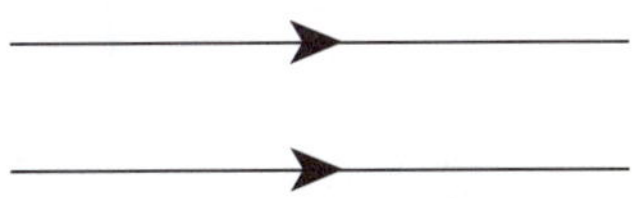

SPACE

## ANGLES

An angle is the space formed by two lines or surfaces that meet. Angles are measured in degrees.

The symbol used for degree is °.

There are different types of angles.

### Acute Angle

An angle less than 90°

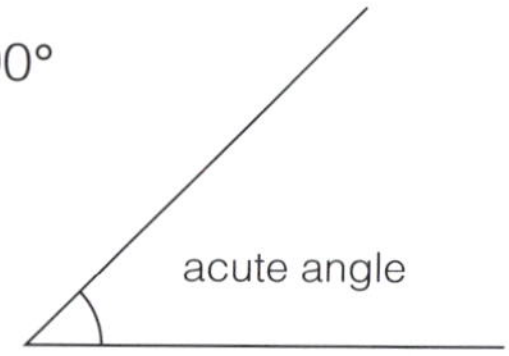


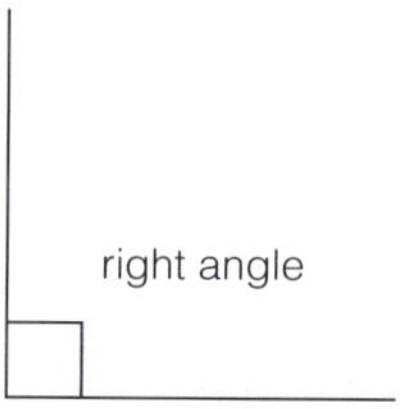


### Right Angle

An angle of exactly 90°

### Obtuse Angle

An angle between 90° and 180°

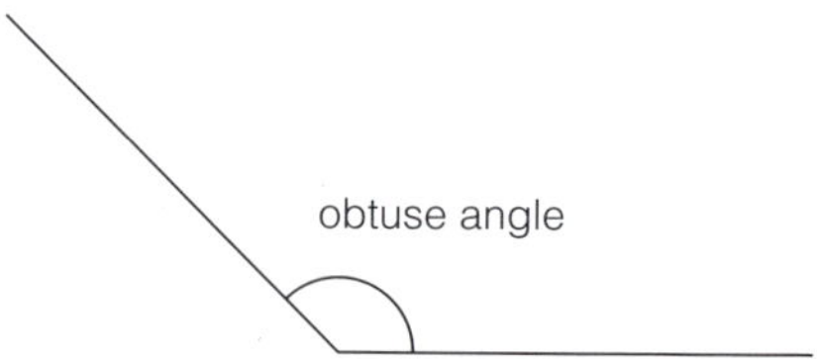


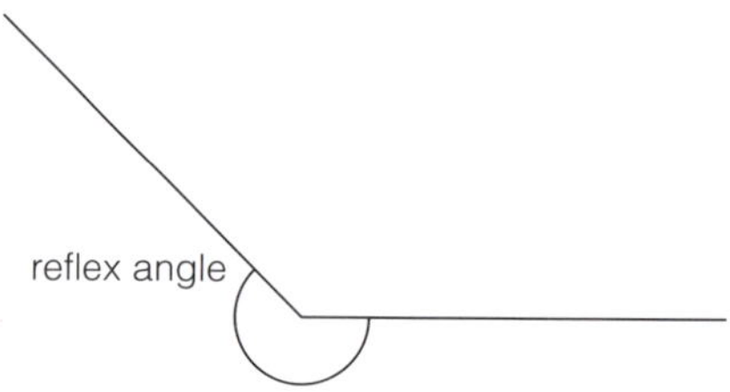


### Reflex Angle

An angle between 180° and 360°

A protractor is a measuring tool used to measure the size of an angle.

## PLANE FIGURES

Plane figures are those shapes that occupy a space on a plane. They may have curved or straight sides.

Plane figures with straight sides are called polygons.

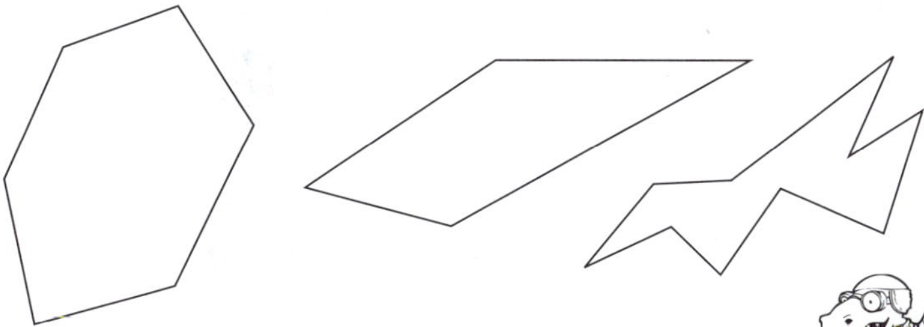

Those polygons with sides of equal length are called regular polygons.

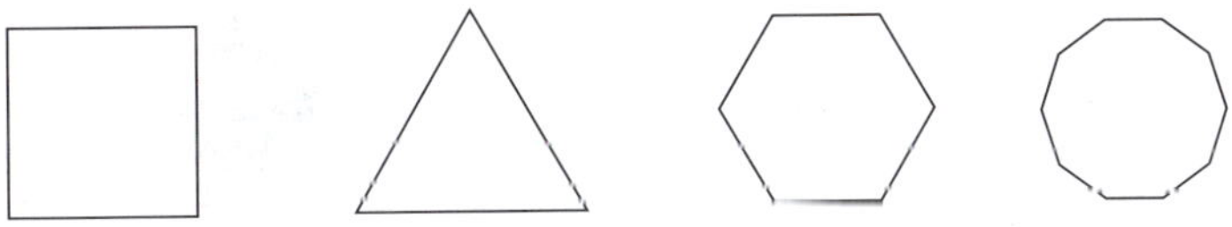

Other plane shapes include circles and ellipses.

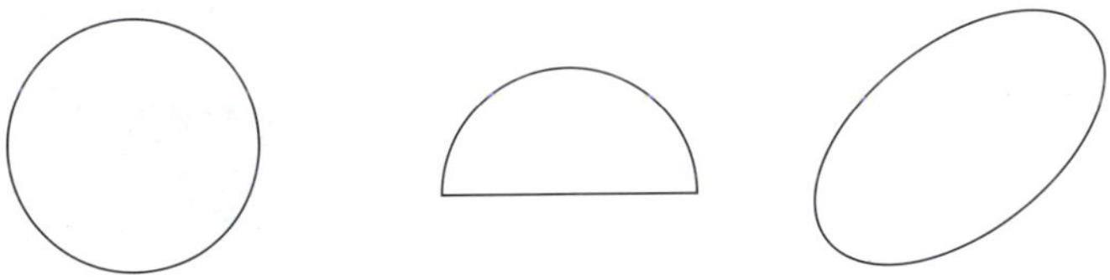

Parts of a circle include the following:

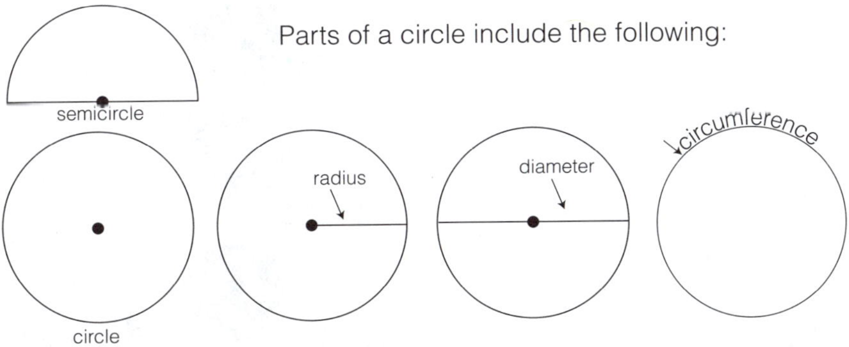

SPACE

## TWO-DIMENSIONAL SHAPES

| Sides/ Angles | Name | Shape |
|---|---|---|
| 3 | triangle |  |
| 4 | square |  |
| 4 | rectangle |  |
| 5 | pentagon |  |
| 6 | hexagon |  |
| 7 | heptagon |  |
| 8 | octagon |  |
| 9 | nonagon |  |
| 10 | decagon |  |

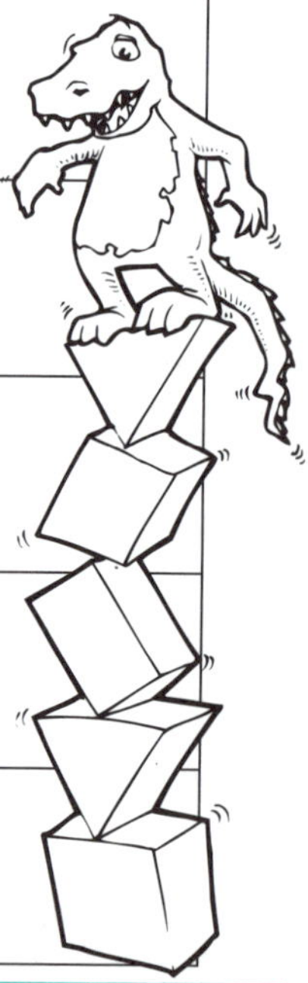

## SOLID FIGURES

Solid figures are also called three-dimensional figures.
A solid figure has length, width and height.

Parts of a solid figure include the following.

Faces: the surface area of a three-dimensional figure.

Edges: the intersection of two faces of a three-dimensional figure.

Vertices: the intersection of two or more edges of a three-dimensional figure.

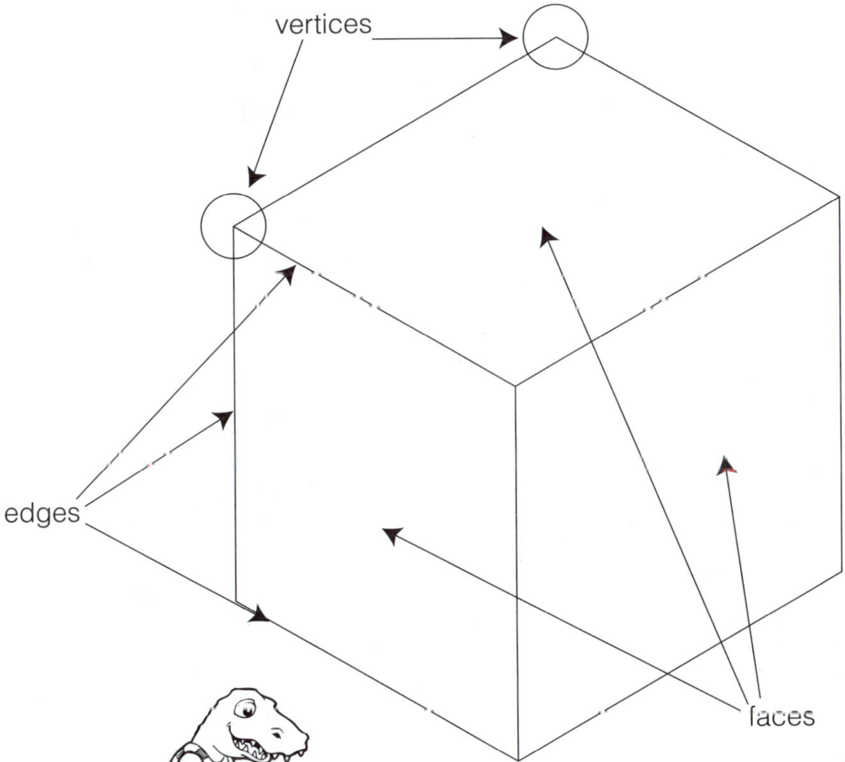

SPACE

## THREE-DIMENSIONAL SHAPES

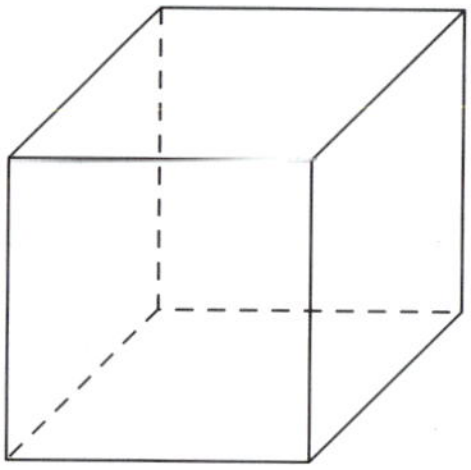

Cube

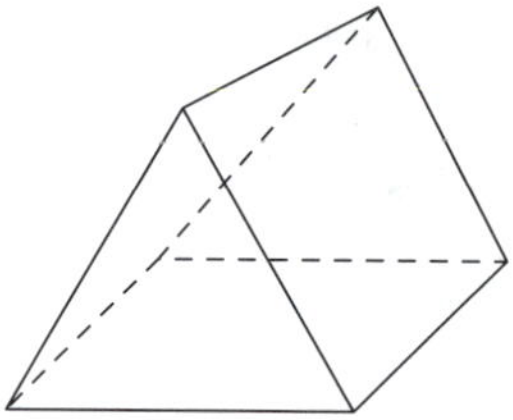

Triangular Prism

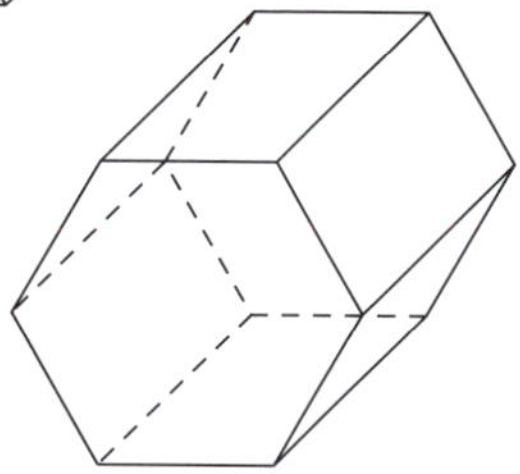

Hexagonal Prism

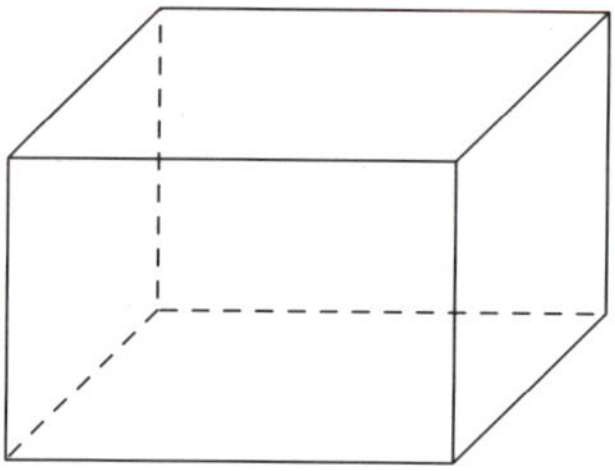

Rectangular Prism

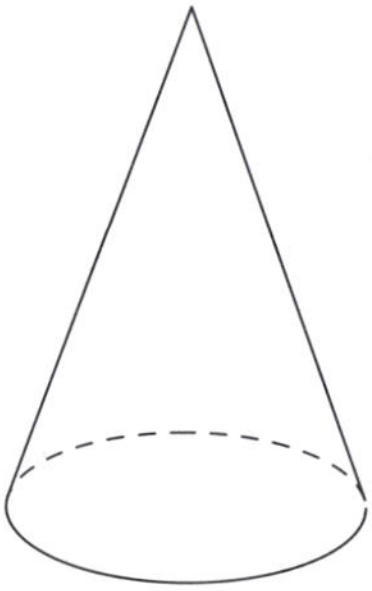

Cone

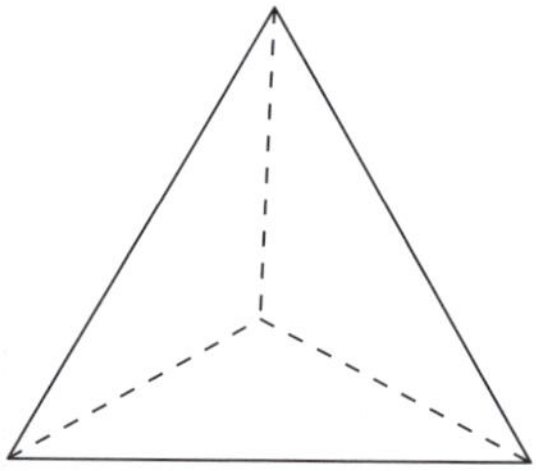

Triangular Pyramid

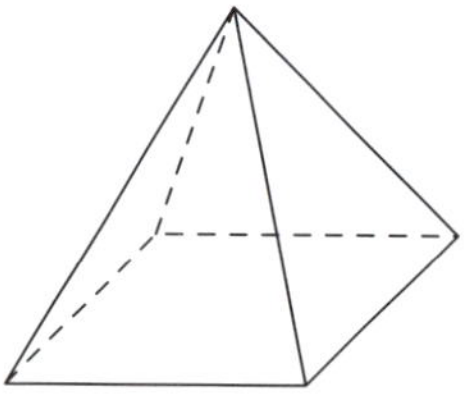

Square Pyramid

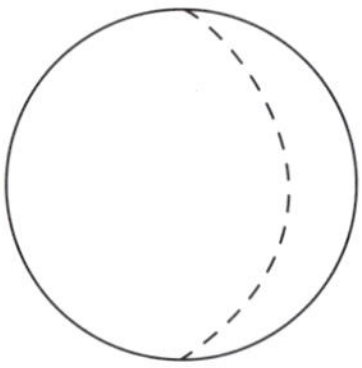

Sphere

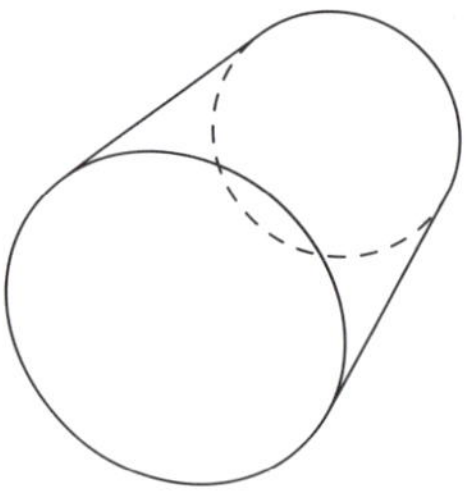

Cylinder

## GRAPHS AND DIAGRAMS

### Bar Graph

A graph which represents information regarding frequency of outcomes using bar lengths.

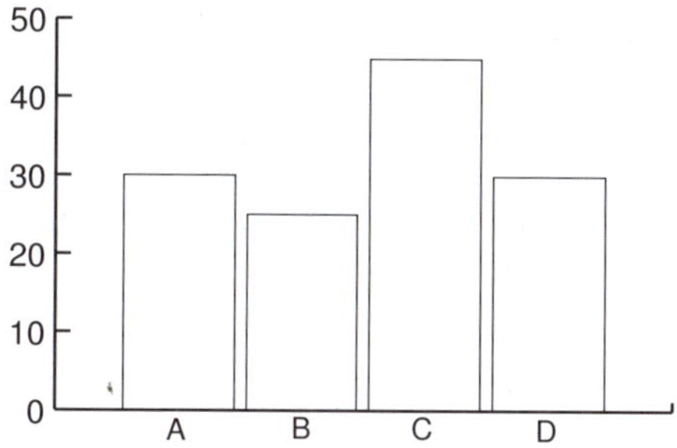


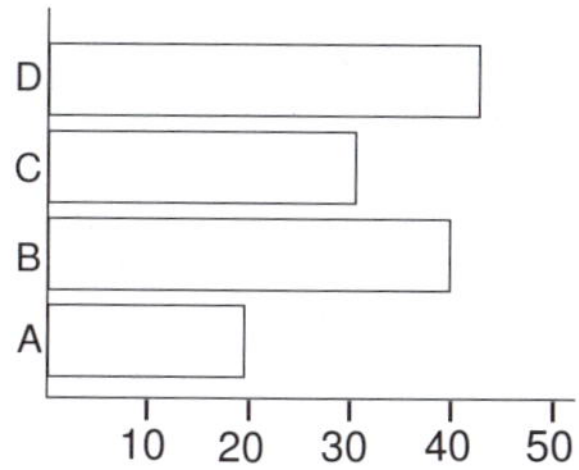


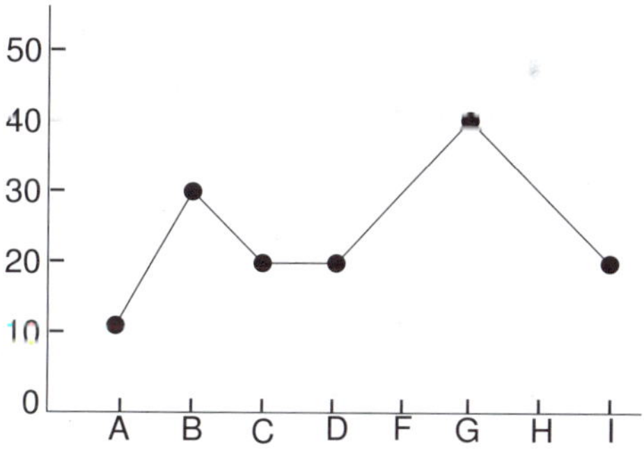


### Line Graph

A graph formed by drawing straight lines which join points representing data.

### Pictograph

A graph in which data is represented by pictures. One picture could represent one unit or many.

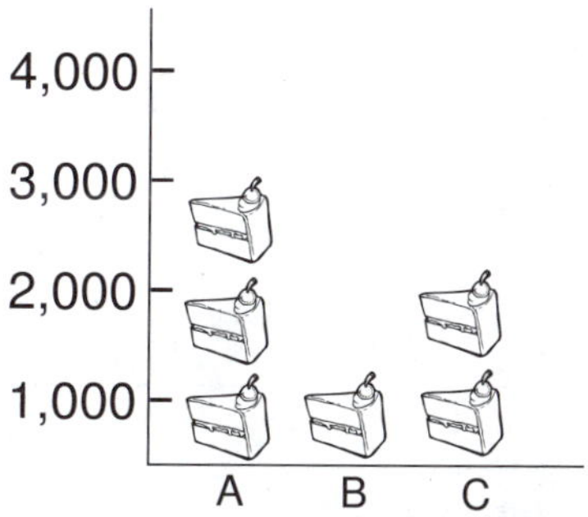

SPACE

## GRAPHS AND DIAGRAMS

### Pie Graphs

A graph in which portions of a circle are used to show a total divided into parts.

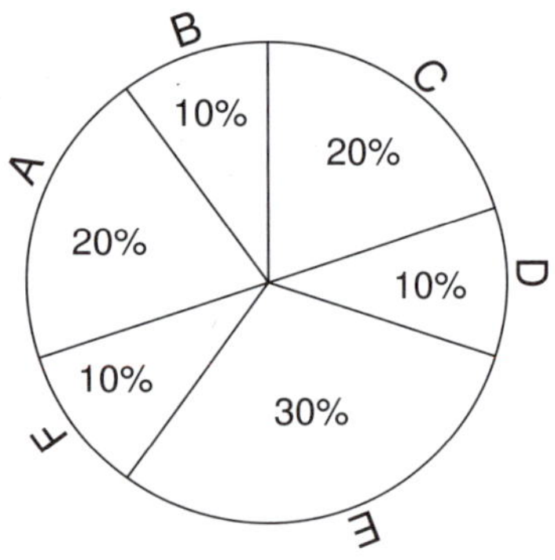


### Tally

This is when a stroke is used to record items.
Tallies generally use a diagonal stroke on the fifth item.

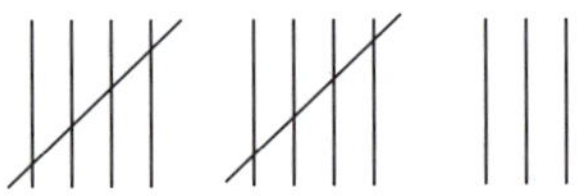 = 13

### Tables

These are used to organize data for a particular purpose.

| Number of children on each team | | | | |
|---|---|---|---|---|
| | **Blue** | **Red** | **Green** | **Gold** |
| **Boys** | 46 | 43 | 49 | 32 |
| **Girls** | 40 | 50 | 35 | 49 |
| **Total** | 86 | 93 | 84 | 81 |

## GRAPHS AND DIAGRAMS

### Carroll Diagram

Named after Lewis Carroll, author, mathematician and logician.
This diagram is useful when recording classification activities.

| | **square** | **not square** |
|---|---|---|
| **blue** | | |
| **not blue** | | |

### Tree Diagrams

These are used for classification activities or to show possible outcomes of chance events.

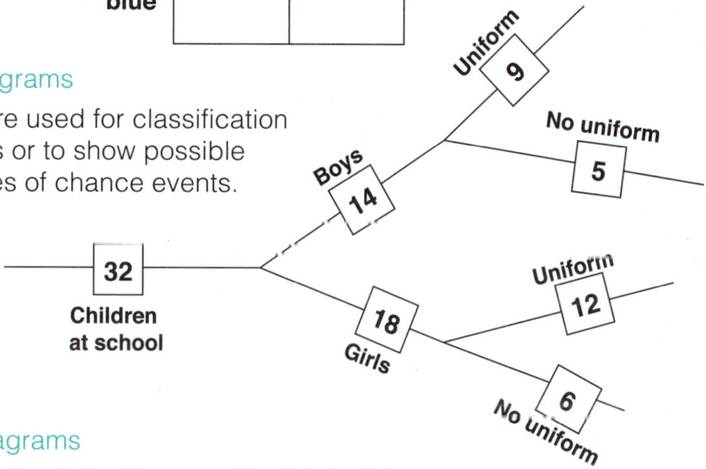


### Venn Diagrams

Named after John Venn, an English logician.
This diagram is used to represent the classification of sets of items.

Counting Numbers 1–20

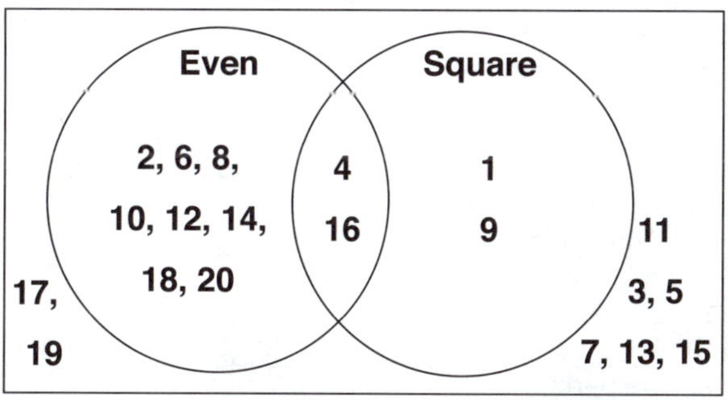

## IMPORTANT MATHEMATICAL TERMS

**abacus:**
a device for calculating; can use beads or counters.

**acute:**
describes an angle between 0° and 90°.

**addend:**
any number which is to be added: 2 + 5 = 7 (2 and 5 are addends).

**adjacent:**
adjoining (as used to describe lines and angles).

**alternate:**
every other one in a sequence.

**angle:**
the number of degrees rotated around a point.

**arbitrary units:**
non-standard units such as hand spans, counters and hand claps.

**arc:**
part of the circumference of a circle.

**area:**
the amount of space within a perimeter (expressed in square units).

**array:**
the arrangement of the units of a number in rows and columns (e.g., 6 = : : : ).

**ascending order:**
the arrangement of numbers from smallest to largest.

**average:**
a number representing a set of numbers (obtained by dividing the total of the numbers by the number of numbers in the set). See "mean."

**axis (of symmetry):**
a line dividing a shape into two symmetrical parts.

**bar graph:**
a diagram representing information by the length of bars.

**base:**
the line or face on which a shape is standing.

**block graph:**
a diagram representing information by the length of blocks.

**capacity:**
the amount of space in the interior of an object (the amount of liquid/air it contains).

**Carroll diagram:**
a problem-solving diagram used in classification activities.

**chord:**
line segment connecting two points on the arc of a circle.

**chronological:**
describes dates arranged in the order in which they occur (e.g., 1920, 1936, 1992).

**circumference:**
the distance around a circle (its perimeter).

**clockwise:**
moving in the same direction as the hands on a clock.

**closed curve:**
a shape with no end points.

**common numeral:**
the symbol commonly used to represent a mathematical expression (e.g., the common numeral for $10^3$ is 1,000).

**composite number:**
a number with more than two factors.

**concentric circles:**
circles which have the same center point.

## IMPORTANT MATHEMATICAL TERMS

**congruent:**
congruent shapes are the same shape and size (equal).

**consecutive:**
consecutive numbers follow in order without interruption (e.g., 11, 12, 13).

**coordinates:**
numbers used to locate a point on a grid.

**counter-clockwise:**
moving in the opposite direction from the hands on a clock.

**cubic:**
a cubic number is the answer when the square of a number is multiplied by the number itself.

**cylinder:**
a circular prism with two congruent, parallel and circular end faces.

**denominator:**
the number below the line in a fraction.

**descending order:**
the arrangement of numbers from largest to smallest.

**diagonal:**
a straight line connecting two non-adjacent vertices (corners) of a polygon.

**diameter:**
a straight line connecting two points on the circumference of a circle and passing through the circle's center.

**difference:**
by how much a number is bigger or smaller than another.

**digit:**
any whole number from 0 to 9 (inclusive).

**digital clock:**
a clock that shows the time by using numbers rather than hands.

**dimensions:**
the measurements of a shape (e.g., length, width, height).

**dodecagon:**
a polygon with 12 sides and 12 angles.

**dodecahedron:**
a three-dimensional shape with 12 faces.

**doubling:**
multiplying a number or shape by two.

**edge:**
the intersection of two faces of a three-dimensional object.

**equation:**
a statement of equality between two expressions (e.g., 3 x 4 = 6 + 6).

**equilateral triangle:**
a triangle with congruent (equal) sides and angles.

**equivalent:**
having the same value.

**even number:**
a positive (e.g., 8) or negative (e.g., –6) number exactly divisible by two.

**face:**
a plane surface of a three-dimensional object.

**face value:**
the numeral itself despite its position in a number (e.g., the face value of the 5 in 3,520 is 5).

**factor:**
a number which will divide evenly into another.

## IMPORTANT MATHEMATICAL TERMS

**fraction:**
an amount expressed in terms of a numerator and a denominator (usually part of a whole).

**heptagon:**
a polygon with seven sides and seven angles.

**hexagon:**
a polygon with six sides and six angles.

**hexagonal:**
having the shape of a hexagon.

**hexahedron:**
a three-dimensional shape with six faces.

**horizontal:**
describes a line or plane parallel to the horizon.

**improper fraction:**
a fraction whose numerator is equal to or greater than its denominator.

**index notation:**
a short way of writing large numbers by multiplication of repeated factors (e.g., $3^2$).

**integer:**
a negative or positive whole number (e.g., …–2, –1, 0, 1, 2…).

**intersection:**
the point or line where two lines or two faces meet or, in set theory, those elements common to two or more sets.

**irregular shapes:**
shapes which do not have all congruent sides and all congruent angles.

**isosceles triangle:**
a triangle with two congruent sides and two congruent angles.

**line graph:**
a diagram using straight lines to join points representing certain information.

**line segment:**
part of a line between two given points.

**mass:**
the amount of matter in an object (its weight on Earth).

**mean (arithmetic mean):**
the average of a set of numbers.

**median:**
in statistics, the middle measurement when information is arranged in order of size (e.g., 5 is the median of 2, 3, 5, 10, 13). Where there is no middle score, an average of the two central scores is taken.

**mixed numeral:**
a symbol representing a whole number and a fraction (e.g., $2\,^{3}/_{8}$).

**modal score:**
in statistics, the measurement that occurs most often (e.g., the modal score of 2, 7, 4, 4, 3, 4, 9, 4 is 4).

**multiples:**
the multiples of a number are those numbers which a given number will divide into exactly (e.g., some multiples of 3 are 3, 6, 9 and 12).

**net:**
a flat pattern that can be folded to make a three-dimensional model of a shape.

**network:**
a system of lines (paths) and nodes (points representing intersections).

## *IMPORTANT MATHEMATICAL TERMS*

**numeral:**
a symbol used to represent a number (e.g., 5 and V are symbols representing the number 5).

**numerator:**
the number above the line in a fraction.

**obtuse angle:**
an angle between 90° and 180°.

**octagon:**
a polygon with eight sides and eight angles.

**octahedron:**
a polyhedron (three-dimensional shape) with eight faces.

**odd number:**
a number that leaves a remainder of 1 when divided by 2.

**parallel lines:**
lines with no common points and always the same distance apart.

**parallelogram:**
a four-sided polygon with opposite sides equal and parallel and containing no right angles.

**path:**
a line connecting nodes (points) in a network.

**pentagon:**
a polygon with five sides and five angles.

**pentomino:**
a plane shape made of five congruent squares connected to each other by at least one common side.

**percentage:**
a quantity expressed in hundredths (%).

**perfect number:**
a number which is the sum of its factors excluding itself (e.g., 6 = 1 + 2 + 3).

**perimeter:**
the length of the distance around the boundary of a shape.

**perpendicular line:**
a line at right angles to another line or plane.

**pie graph:**
a circular graph in which sectors of a circle are used to show information.

**place value:**
indicates the position of a numeral (e.g., the place value of the 3 in the number 357 is hundreds).

**polygon:**
a two-dimensional shape with three or more straight lines.

**polyhedron (plural – polyhedra):**
a three-dimensional shape with plane faces.

**prime factor:**
a prime number that will divide exactly into another number (e.g., 2 and 3 are prime factors of 6).

**prime number:**
a number with only two factors, 1 and itself.

**prism:**
a three-dimensional shape with at least one pair of opposite faces which are congruent and parallel.

**product:**
the result when two or more numbers are multiplied (e.g., the product of 2, 3 and 4 is 2 x 3 x 4 = 24).

**proper fraction (vulgar or common fraction):**
a fraction in which the numerator is less than the denominator.

## IMPORTANT MATHEMATICAL TERMS

**property:**
an attribute of a two-dimensional or three-dimensional shape.

**protractor:**
a semi-circular or circular instrument for measuring angles.

**quadrant:**
a quarter of the area of a circle which also contains a right angle.

**quadrilateral:**
any four-sided polygon.

**radius:**
a line joining the center of a circle to a point on the circle's circumference.

**rectangle:**
a quadrilateral with opposite sides equal and parallel and containing four right angles (a square is a rectangle).

**reflex angle:**
an angle greater than 180°.

**region:**
the interior area enclosed by a perimeter (the area outside the perimeter is the exterior region).

**regular shape:**
a polygon is regular if all its sides and angles are congruent (opposite – irregular).

**rhombus:**
a parallelogram with congruent sides and containing no right angles (a "diamond" shape).

**right angle:**
an angle containing 90°.

**right-angled triangle:**
a triangle containing one right angle (90°).

**rotating:**
turning in a clockwise or counter-clockwise direction.

**scalene triangle:**
a triangle with sides of different length and three different interior angles.

**sector:**
the part of a circle bounded by two radii and the included arc.

**segment:**
the part of a circle bounded by a chord and an arc.

**semi-circle:**
half of a circle (the area bounded by a diameter and an arc).

**simplify:**
to change to simpler terms (e.g., $^{45}/_{60}$ to $^{3}/_{4}$).

**solid shape:**
a three-dimensional shape (with length, width, height).

**sphere:**
a three-dimensional shape comprising a set of points, each point being equidistant from its center (e.g., a ball).

**square number:**
a number whose units can be arranged into a square (e.g., 4 ∷).

**square pyramid:**
a three-dimensional shape with a square base and four sloping triangular sides meeting at a common vertex.

**squared:**
a number squared is multiplied by itself.

**sum:**
the result when two or more numbers are added.

## IMPORTANT MATHEMATICAL TERMS

**supplementary angles:**
two angles which, when added, total 180°.

**surface area:**
the total area of all faces of a three-dimensional shape (measured in square units).

**symmetrical:**
a shape is symmetrical if it is identical on either side of a line dividing it into two parts.

**tally:**
a record of items using vertical and oblique lines to represent each item.

**tessellating shapes:**
shapes which cover an area with no gaps between them (e.g., squares).

**tetrahedron:**
a polyhedron with four faces (e.g., triangular pyramid).

**tetromino:**
a plane shape made of four congruent squares connected to each other by at least one common side.

**three-dimensional shape:**
a shape with length, width and height.

**total:**
the result when two or more numbers are added.

**trapezium:**
a quadrilateral plane figure of which no two sides are parallel.

**trapezoid:**
a quadrilateral plane figure in which only one pair of opposite sides is parallel.

**traversable:**
a network is traversable if all paths can be traced over without going over the same path twice.

**triangular number:**
a number whose units can be arranged into a triangle (e.g., 6).

**twin primes:**
prime numbers separated by a composite number (e.g., **3**, 4, **5**).

**Venn diagram:**
a diagram which shows sets and their relationships.

**vertex (plural – vertices):**
the point at which two or more line segments or two or more edges of a polyhedron meet.

**vertical line:**
a line which is at right angles to a horizontal line.

**volume:**
the amount of space taken up by an object.

**Other words of interest.**

______________________________

______________________________

______________________________

______________________________

______________________________

______________________________

## 100 DIFFICULT WORDS TO SPELL

ache
again
answer
always
any
among
believe
blue
break
business
been
beginning
busy
built
buy
can't
choose
color
coming
cough
could
country
dear
doctor
does
done
don't
early

easy
enough
every
February
forty
friend
grammar
guess
half
having
hear
heard
here
hoarse
hour
instead
just
knew
know
laid
loose
lose
many
making
meant
minute
much
none

often
once
piece
raise
read
ready
said
says
seems
separate
shoes
since
some
straight
sure
sugar
tear
their
there
they
though
through
tired
tonight

too
trouble
truly
Tuesday
two
used
very
wear
Wednesday
week
where
whether
which
whole
women
won't
would
write
writing
wrote

## DIFFICULT WORDS I NEED TO LEARN

## SPELLING RULES

1. Change the final "y" to "i" before adding any suffix, except with "ing" or when "y" follows a vowel.

   e.g., carry–carried; carry–carrying

2. A short vowel sound should always be followed by two consonants when adding "y," "ed," "ing," or "er."

   e.g., run–running; jump–jumper

3. When the words "all," "full," "fill" and "till" are used as suffixes and prefixes, one "l" is dropped.

   e.g., already; fulfil

4. When a word ends in an "e" the "e" is usually dropped when a suffix beginning with a vowel is added.

   e.g., bake + ing = baking

5. When a word ends in an "e" the "e" is usually kept when a suffix beginning with a consonant is added. Usually the "e" is dropped when adding a suffix beginning with a vowel.

   e.g., price + less = priceless; love + able = lovable

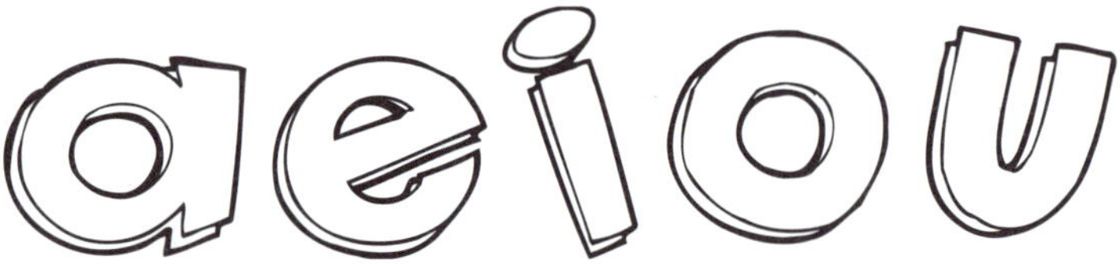


6. The prefixes "dis–" and "mis–" never drop their "s," even when added to a word beginning with "s."

   e.g., dis–ability; dis–satisfied

7. Words ending in "–ce" and "–ge" drop the "–e" when "–ing" is added but retain the "–e" when "–able" and "–ous" are added.

   e.g., notice; noticing; noticeable

8. Short "a," "e," "i," "o" and "u," are followed by "ck" (in one syllable words).

   e.g., duck; pluck

9. "k" goes in front of "e" and "i"; "c" goes in front of "a," "o" and "u."

   e.g., kelp, kick; cat, cot, cut

## *SPELLING RULES*

10. "k," "g" and "p" are silent at the beginning of a word if followed by a consonant (usually "n").

    e.g., knee; gnaw; pneumonia

11. Adjectives ending in "–le" after a consonant form adverbs by dropping the "e" and adding a "y."

    e.g., gentle; gentl–y

12. "i" before "e" except after "c" with the sound of "long e" (exception "seize").

    e.g., brief; receive

13. Most words of two or more syllables ending in "l" have only one final "l."

    e.g., parallel; initial

14. Words which end in a single consonant preceded by a long vowel do not double the consonant when a suffix beginning with a vowel is added.

    e.g., beat; beaten; beating; beater

15. Words ending in a single consonant other than "l," but not accented on the last syllable, do not double the consonant when a suffix is added.

    e.g., gallop; galloper; galloping

16. Words which end in a single consonant preceded by a short, accented vowel sound double the consonant when a suffix beginning with a vowel is added.

    e.g., bat; batted; batting; batter

17. *Syllables*

    A syllable is a unit of sound. Words are made up of syllables. A word contains a syllable for every pronounced vowel.

    e.g., run = 1 syllable
    run/ning = 2 syllables

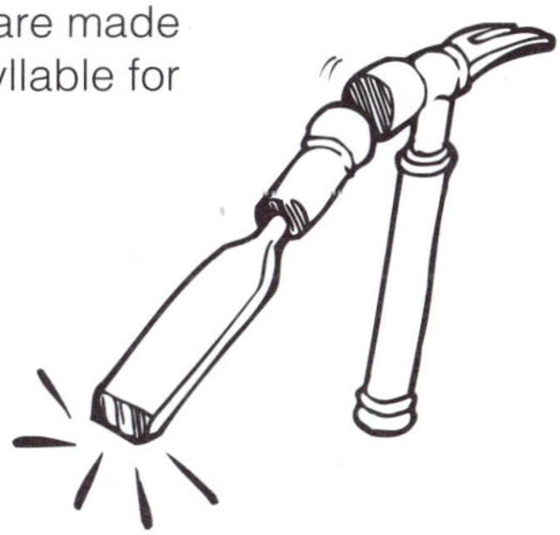

## *PREFIXES*

A prefix is a word part added to the beginning of a word. It can change the meaning of that word.

| Prefix | Meaning | Example |
|---|---|---|
| a | on, at, up, out, to | ashore, asleep, ahead |
| a–, ab– | away from | absent |
| anti– | against | antiseptic |
| auto– | self | autobiography |
| bi– | two | bicycle |
| circum– | around | circumference |
| co–, com–, con– | together, with | companion, conflict |
| contra | against | contrary, contradiction |
| de– | down, away | descend |
| dif–, dis– | apart from, not | different, disagree |
| e–, ex–, out– | out, or out of | export, emigrate, outside |
| extra– | outside, beyond | extraordinary |
| fore– | in front, before | forehand, forehead |
| il– | not | illegal, illegible |
| im–, in– | not | impossible, independent |
| im–, in– | into, in | inspect, inland, immerse |
| inter– | between, among | interact, intersection |
| ir– | not | irregular, irresponsible |
| mid– | middle | midday, midnight |
| mis– | wrong | misfortune, misunderstand |
| non– | not | nonsense |
| over– | over, beyond, too much | overgrown, overcharge |
| post– | after | postpone, postscript |
| pre– | before | precaution, prefix |
| re– | again, back | redo, replace, return |
| pro– | for | proceed, produce |
| super– | over | supermarket, supersonic |
| sub– | under | submarine |
| tele– | far away | telegram, television |
| trans– | across | transport |
| un– | not | unfair, uneven, unselfish |
| under– | under, beneath | underline, underneath |
| after– | following | afternoon, afterthought |
| down– | downwards | downfall, downpour |
| en– | make | enclose, enrich |

## SUFFIXES

A suffix is a word part added to the end of a word. Sometimes when suffixes are added, the spelling of the base word changes e.g. happy – happily.

| Suffix | Meaning | Example |
|---|---|---|
| –ive | having nature of | active |
| –ist | one who | artist, journalist |
| –ish | like | whitish, foolish, childish |
| –less | without | worthless, helpless |
| –ly | in the manner of | happily, sadly, quietly |
| –ment | resulting state, action | treatment, employment |
| –most | most (in degree) | topmost, northernmost |
| –ness | quality or state of being | weakness, sickness |
| –or | person who | actor |
| –ous | state or condition | nervous, dangerous |
| –ship | state of being | hardship, friendship |
| –ure | act, process | pleasure, adventure |
| –ion | act, process, state | action, education |
| –able | tending to, able to | enjoyable, movable |
| –age | state of being, place of, result of | wastage, shrinkage, orphanage |
| –al | relating to, belonging to | electrical, residential |
| –ance, –ence | state of being | excellence, disappearance |
| –ary, ery | that which, place where | bakery, parliamentary |
| –en | having nature of | fallen, broken, golden |
| –en | to make or become | widen, whiten, flatten |
| –er | one who, that which | worker, teacher |
| –er | more (in degree) | taller, faster, smarter |
| –est | most (in degree) | cleanest, loudest |
| –ful | full of | cheerful, helpful |
| –fy | make or form into | identify, notify |
| hood | state of rank | childhood, neighborhood |
| –ible | tending to, able to | sensible, possible |
| –ic | like, belonging to | athletic, artistic |
| –an | pertaining to | comedian, Australian |
| –ese | belonging to | Chinese, Japanese |
| –ling | small, little | duckling, gosling |
| –ate | to make | donate, separate |

## HOMONYMS

Homophones are a type of homonym. Homophones are words that sound the same but are spelled differently.

| | | |
|---|---|---|
| air/heir | meet/meat | threw/through |
| allowed/aloud | metal/mettle | to/two/too |
| alter/altar | muscle/mussel | waist/waste |
| bear/bare | need/knead | wear/where |
| blew/blue | night/knight | weather/whether |
| board/bored | no/know | week/weak |
| bow/bough | none/nun | weight/wait |
| bury/berry | our/hour | wheel/weal |
| by/buy | paced/paste | which/witch |
| carat/carrot | peace/piece | wholly/holy |
| cent/scent/sent | pear/pair/pare | won/one |
| cheep/cheap | plain/plane | wood/would |
| chews/choose | pore/pour | wrap/rap |
| creek/creak | road/rowed/rode | write/right |
| current/currant | rote/wrote | |
| die/dye | sail/sale | |
| eight/ate | scene/seen | |
| faint/feint | see/sea | |
| fair/fare | seam/seem | |
| feet/feat | stare/stair | |
| guest/guessed | steak/stake | |
| hair/hare | steel/steal | |
| herd/heard | straight/strait | |
| here/hear | sun/son | |
| hole/whole | tail/tale | |
| horse/hoarse | teem/team | |
| key/quay | their/there/they're | |

Other homonyms are called homographs. Homographs are spelled the same and sound the same, but have different meanings.

*For example:*
bear and bear, rock and rock, chip and chip, nail and nail.

## CONTRACTIONS

Contractions are found in both written and spoken English. They are a shortened way of writing or speaking two words. An apostrophe of contraction – ' – is used to show where a letter or letters are left out when joining two words to make a contraction.

**n't for not**

wasn't
doesn't
aren't
hadn't
can't
won't
wouldn't
shouldn't
didn't
don't
isn't
haven't
weren't
couldn't

**'s for is**

it's
there's
he's
where's
who's
she's

**'ll for will**

we'll
she'll
I'll
you'll
they'll
who'll

**'re for are**

we're
you're
they're

**'ve for have**

we've
you've
they've

**'d for would**

we'd
I'd
they'd
you'd
he'd

**'m for am**

I'm

## *ROOT WORDS*

Root words are words from another language which have been used to form words in the English language.

| Root | Meaning |
|---|---|
| acro– | top |
| aeq. | equal |
| aero | air |
| alter | other |
| alto | high |
| amor | love |
| amphi– | on all sides |
| –androus | male |
| anima | breath, life |
| aniso– | unlike |
| anthropo– | human |
| apia | bee |
| aqua | water |
| audio– | hear |
| auto– | self |
| avi– | bird |
| baro– | weight |
| bene– | good, well |
| bi | twice, two |
| biblio– | book |
| bio– | life |
| brevi– | short |
| cardio– | heart |
| capere | to take hold |
| caput | head |
| centum | hundred |
| –chrome | color |
| chrono– | time |
| –cide | killer |
| circum– | around |
| civis | citizen |
| –claudere | shut, close |
| cor | heart |
| cosmeo | world |
| counter– | against |
| cranium– | skull |
| credo | I believe |

| Root | Meaning |
|---|---|
| crypto– | hidden |
| crystallo– | crystal |
| cursus | running |
| cycl– | cycle |
| cyto– | cell |
| cyst– | cyst |
| –cyte | cell |
| deca– | ten |
| demo– | people |
| demon– | demon |
| denti– | tooth |
| –derm | skin |
| ducere | to lead |
| duo– | two |
| dys– | bad |
| equi | equal |
| endo– | within, inside |
| entero– | intestine |
| ethno– | race, tribe |
| factor | do |
| fertilis | to make fruitful |
| finis | end |
| fluor | flow |
| gastro– | stomach |
| geo– | earth |
| –gerous | producing |
| –gram | letter, writing |
| –graph | write |
| gyn– | woman |
| gyro– | circle |
| hetero– | other |
| holo– | whole |
| homo– | same |
| ideo– | idea |
| manus | hand |
| mille | thousand |
| mini– | small |

## ROOT WORDS

| Root | Meaning |
|---|---|
| minuere | to lessen |
| mittere | to send |
| mono– | one, alone |
| –morph | form |
| mors | death |
| movere | move |
| multi– | many |
| my– | muscle |
| myc– | fungus |
| myth– | myth |
| myx– | slimy |
| natus | born |
| navis | ship |
| nector | dead |
| neo– | new |
| nepho– | cloud |
| nephro– | kidney |
| nocti– | night |
| novem | nine |
| octa | eight |
| omni– | all |
| ortho– | straight, right |
| paed– | child |
| pan– | all |
| patho– | suffering |
| pedis | foot |
| pendere | hang |
| penta– | five |
| –phasia | disordered speech |
| phen | used in chemical terms |
| benz– | derivative |
| –philia | lover of |
| phlebo– | vein |
| –phobia | fear |
| phon– | sound, voice |
| –phorous | bearing, having |

| Root | Meaning |
|---|---|
| phosp– | phosphorus |
| photo– | light |
| physio– | physical |
| phyto– | plant |
| –plasty | formation |
| poly– | many |
| portare | carry, bear |
| prima | first |
| project | throw |
| proto– | first |
| pseudo– | false |
| psych– | mind |
| ptero– | wing |
| pyo– | pus |
| quadri– | four |
| quinque– | five |
| recti– | set upright |
| scrib | write |
| –sect | cut |
| septum | seven |
| sex– | six |
| Sino– | Chinese |
| spectro– | look |
| –sphere | sphere |
| spiro– | spiral |
| –stat | stationary |
| statua | statue |
| tele– | afar, end |
| terra | earth |
| tractum | trace |
| tetra | four |
| tri– | three |
| uni– | one |
| video | see |
| vox | voice |

## SHORTENED FORMS AND ACRONYMS

### A

| | |
|---|---|
| abb. or abbrev. | abbreviation |
| AD | in the year of our Lord (anno Domini) |
| AIDS | acquired immune deficiency syndrome |
| a.m. | before noon (ante meridiem) |
| anon. | anonymous |
| ans. | answer |
| approx. | approximately |
| ASAP | as soon as possible |
| Aug. | August |
| Aust. | Australia |
| Ave. | Avenue |

### B

| | |
|---|---|
| BBQ | barbecue |
| BC | Before Christ |
| Blvd. | Boulevard |
| bros. | brothers |
| BYO | bring your own |

### C

| | |
|---|---|
| C | Cape; Celsius; Centigrade; century |
| C/– | care of |
| ¢ | cent |
| Capt. | Captain |
| CD | compact disc |
| Co. | company, county |
| COD | cash on delivery (also C.O.D.) |
| Corp. | corporation, Corporal |
| Cres. | Crescent |
| CST | Central Standard Time |

### D

| | |
|---|---|
| dec. | deceased |
| Dec. | December |
| DOB | date of birth |
| doz. | dozen |
| Dr. | Doctor; Drive |

### E

| | |
|---|---|
| E | east; eastern |
| e.g. | for example (exempli gratia) |
| equiv. | equivalent |
| ESP | extrasensory perception |
| Esq. | Esquire |
| est. | established |
| EST | Eastern Standard Time |
| etc. | and so on (et cetera) |

### F

| | |
|---|---|
| F | Fahrenheit |
| Feb. | February |
| fem. | feminine |
| Fr. | Father |
| Fri. | Friday |
| fwd. | forward |
| f.w.d. | four-wheel drive/front-wheel drive |

### G

| | |
|---|---|
| GM | General Manager/General Motors |
| GMT | Greenwich Mean Time |
| Govt. | Government |

### H

| | |
|---|---|
| hdqrs. | headquarters |
| HO | Head Office |
| Hon. | honorary; honorable |
| HQ | Headquarters |

### I

| | |
|---|---|
| ID | identification |
| i.e. | that is (id est) |
| IOU | I owe you |
| ISBN | International Standard Book Number |

### J

| | |
|---|---|
| Jan. | January |
| Jun. | June |
| JP | Justice of the Peace |
| jr. | junior |
| Jul. | July |

### K

| | |
|---|---|
| km | kilometer |
| km/h | kilometers per hour |

## *SHORTENED FORMS AND ACRONYMS*

### L

| | |
|---|---|
| lat. | latitude |
| LPG | liquified petroleum gas |

### M

| | |
|---|---|
| Mar. | March |
| masc. | masculine |
| math | mathematics |
| Messrs | plural of Mr. |
| mill. | million(s) |
| min. | minimum |
| misc. | miscellaneous |
| Mon. | Monday |
| MP | Military Police |
| Mr. | Mister |
| Mrs. | Mistress (missus) |
| Mt. | mount; mountain |

### N

| | |
|---|---|
| N | north |
| n/a | not applicable |
| NB | note well (nota bene) |
| NE | northeast |
| No. | number |
| NW | northwest |

### O

| | |
|---|---|
| Oct. | October |
| opp. | opposite |

### P

| | |
|---|---|
| p.a. | per annum |
| p.c. | percent |
| phys. ed. | physical education |
| Pk. | Park |
| Pl. | Place |
| p.m. | afternoon (post meridiem) |
| PO | Post Office |
| POW | prisoner of war |
| PR | public relations |
| Prof. | Professor |
| PS | postscript |

### Q

| | |
|---|---|
| qtr. | quarter |

### R

| | |
|---|---|
| r. | radius |
| Rd. | Road |
| recd. | received |
| Rev. | Reverend |
| r.p.m. | revolutions per minute |
| RSVP | please reply (répondez s'il vous plaît) |

### S

| | |
|---|---|
| S | south |
| Sat. | Saturday |
| SE | southeast |
| Sept. | September |
| St. | Saint; street |
| SPCA | Society for the Prevention of Cruelty to Animals |
| Sun. | Sunday |
| SW | southwest |

### T

| | |
|---|---|
| tbsp. | tablespoon |
| temp. | temperature |
| Thurs. | Thursday |
| tsp. | teaspoon |
| Tues. | Tuesday |

### U

| | |
|---|---|
| UN | United Nations |
| UV | ultraviolet |

### V/W

| | |
|---|---|
| vs | versus |
| W | west |
| Wed. | Wednesday |
| wk. | week |

## PARTS OF SPEECH

**adjective:**
a word used to describe a noun or a pronoun.
e.g., *adventurous* child

**adverb:**
a word used to say when, how, or where about a verb.
e.g., spoke *loudly*

**antonyms:**
a word that is opposite in meaning to another word.
e.g., *good/bad* are antonyms

**conjunction:**
is a word used to join other words.
e.g., The boy and the girl *who* won gold!

**homonym:**
words that sound the same. Homonyms can be either:

- homographs which are words that sound the same, have the same spelling, yet have different meanings e.g., *jam/jam*; or
- homophones which are words which sound the same yet have different spellings. e.g., *aloud/allowed*

**noun:**
a word that says the name of a person, place, or thing.
e.g., *girl, home, string*

A special type of noun is a proper noun which is the name of a particular person, place, or thing. All proper nouns have capital letters to begin the word.
e.g., *Mary, Boston, January*

**preposition:**
a word that shows the connection between a noun and other words in a sentence.
e.g., *beside* the road

**pronoun:**
a word used in the place of a noun.
e.g., *it, him, her, I, them*

**synonyms:**
words that are similar in meaning.
e.g., *big/large*

**verb:**
a word that shows action, "doing" word.
e.g., *speak, drive, walk, look*

## LANGUAGE TERMS

**acronym:**
a word formed by the first letters of other words.
e.g., radar – radio detection and ranging

**alliteration:**
the use of the same sound at the beginning of a series of words.
e.g., ten terrific turtles

**anagram:**
a new word made from the letters of another word.
e.g., tablet – battle

**analogy:**
a likeness between things.
e.g., between a heart and a pump.

**ballad:**
a narrative poem often including a lot of action.

**cinquain:**
a poem of five lines
1st line – a noun
2nd line – two adjectives describing the noun in line one.
3rd line – three words describing the noun in an active form.
4th line – four words telling what the noun does.
5th line – repeats the first line.

**colloquialism:**
Colloquialisms are expressions used in common conversation. They usually sum up a situation and are particular to a certain area.
e.g., hit the nail on the head – to be right

**fable:**
a short tale that teaches a moral and often portrays animals as characters.

**fairytale:**
stories told of amazing happenings. Characters often have special powers and stories relate to times of the past.

**folktale:**
traditional stories, poems and songs passed down from generation to generation.

**haiku:**
A Japanese verse that succinctly describes an aspect of nature in 17 syllables. It has three lines: line 1 – 5 syllables, line 2 – 7 syllables, line 3 – 5 syllables.

**legend:**
folkstories of great significance handed down through generations.

**limerick:**
a five-line nonsense poem where the first, second and fifth lines rhyme, while the third and fourth lines rhyme.

**metaphor:**
a term or phrase is used to compare with something which it is not literally suitable.
e.g., a mighty oak tree is our father

**myth:**
an ancient story made up to explain the existence of a thing or event.

**onomatopoeia:**
the formation of a word by imitating an associated sound.
e.g., hiccup

**proverb:**
a group of words that present a wise thought.
e.g., many hands make light work

**simile:**
comparison between two objects using the words *like* or *as*.
e.g., as cold as ice

**tautologies:**
unnecessary repetition of a message in a sentence using different words.
e.g., he *struggled* upstream with *great effort*

## COLLECTIVE NOUNS

| | | |
|---|---|---|
| anthology | of | stories |
| archipelago | of | islands |
| armada, fleet | of | ships |
| army | of | soldiers |
| assembly | of | people |
| bale | of | wool, hay |
| band | of | musicians |
| bench | of | magistrates |
| board | of | directors |
| bouquet | of | flowers |
| brood | of | chicks |
| bunch | of | flowers, bananas |
| chain | of | mountains |
| chest | of | drawers |
| choir | of | singers |
| class | of | students |
| cluster | of | gemstones |
| collection | of | stamps, coins |
| company | of | actors |
| congregation | of | worshipers |
| constellation | of | stars |
| crew | of | sailors |
| fleet | of | ships |
| flight | of | planes |
| flock | of | birds, sheep |
| flotilla | of | ships |
| forest | of | trees |
| gaggle | of | geese |
| galaxy | of | stars |
| grove | of | trees |
| herd | of | cattle |
| hive | of | bees |
| host | of | angels |
| library | of | books |
| litter | of | pups, pigs |
| mob | of | kangaroos, cattle |
| nest | of | eggs |
| orchard | of | fruit trees |
| orchestra | of | musicians |
| pack | of | wolves |
| plague | of | insects, locusts |
| pride | of | lions |
| range | of | mountains |
| regiment | of | soldiers |
| rope | of | pearls |
| school | of | fish |
| sheaf | of | arrows, wheat |
| shoal | of | fish |
| skein | of | wool, geese when in flight |
| string | of | horses, beads, pearls |
| suit | of | clothes |
| suite | of | furniture |
| swarm | of | insects, bees |
| team | of | sportspeople |
| tribe | of | people |
| troop | of | monkeys, scouts |
| troupe | of | performers |
| vineyard | of | grapes |

## PUNCTUATION MARKS

**apostrophe – '**

- of ownership/possession: where an apostrophe is used to replace *of*.
  e.g., The clothes of the boy. The boy's clothes.
- used to show when a letter or letters have been omitted.
  e.g., did not – didn't

**capital letters – C**

used at the beginning of a sentence, as the first letter in a proper noun, in abbreviations and for the word "I."

**colon – :**

marks a natural pause in reading and is used to introduce a list or series.
e.g., For the football game tomorrow bring: a raincoat, a hat, coffee and sandwiches.

**comma – ,**

- *marks a short pause* in reading to help make a sentence make sense.
  e.g., The team struggled, even though they had won previously.
- *divides items* in a sentence.
  e.g., James invited John, Mary, Gary and Julia to his birthday.
- *marks spoken words* in a sentence
- *used to insert information* within a sentence.
  e.g., The wall, which was falling down, was roped off from the children.

**dash – -**

used to indicate a break in a sentence.
e.g., I'm coming – don't wait – right away.

**ellipsis – ...**

marks where there is a break in thought or where words are left out.

**exclamation points – !**

used at the end of a sentence to show strong feelings or emotion.
e.g., Stop, thief!

**hyphen – -**

used to show how words are joined to create meaning.
e.g., soon-to-be bride

**period – .**

used to show the end of a sentence. Also used in some abbreviations.

**question mark – ?**

used at the end of a sentence to indicate a question is being asked.

**quotation marks – " "**

- used to show where direct speech is being used.
- also used to indicate names of books, movies, poems, etc.

**semicolon – ;**

used to separate parts of a sentence where there is a more distinct break than one designated by a comma.

## STATES AND CAPITALS – UNITED STATES OF AMERICA

| State | Capital |
|---|---|
| Alabama | Montgomery |
| Alaska | Juneau |
| Arizona | Phoenix |
| Arkansas | Little Rock |
| California | Sacramento |
| Colorado | Denver |
| Connecticut | Hartford |
| Delaware | Dover |
| Florida | Tallahassee |
| Georgia | Atlanta |
| Hawaii | Honolulu |
| Idaho | Boise |
| Illinois | Springfield |
| Indiana | Indianapolis |
| Iowa | Des Moines |
| Kansas | Topeka |
| Kentucky | Frankfort |
| Louisiana | Baton Rouge |
| Maine | Augusta |
| Maryland | Annapolis |
| Massachusetts | Boston |
| Michigan | Lansing |
| Minnesota | St. Paul |
| Mississippi | Jackson |
| Missouri | Jefferson City |
| Montana | Helena |
| Nebraska | Lincoln |
| Nevada | Carson City |
| New Hampshire | Concord |
| New Jersey | Trenton |
| New Mexico | Santa Fe |
| New York | Albany |
| North Carolina | Raleigh |
| North Dakota | Bismarck |
| Ohio | Columbus |
| Oklahoma | Oklahoma City |
| Oregon | Salem |
| Pennsylvania | Harrisburg |
| Rhode Island | Providence |
| South Carolina | Columbia |
| South Dakota | Pierre |
| Tennessee | Nashville |
| Texas | Austin |
| Utah | Salt Lake City |
| Vermont | Montpelier |
| Virginia | Richmond |
| Washington | Olympia |
| West Virginia | Charleston |
| Wisconsin | Madison |
| Wyoming | Cheyenne |

## THE WORLD

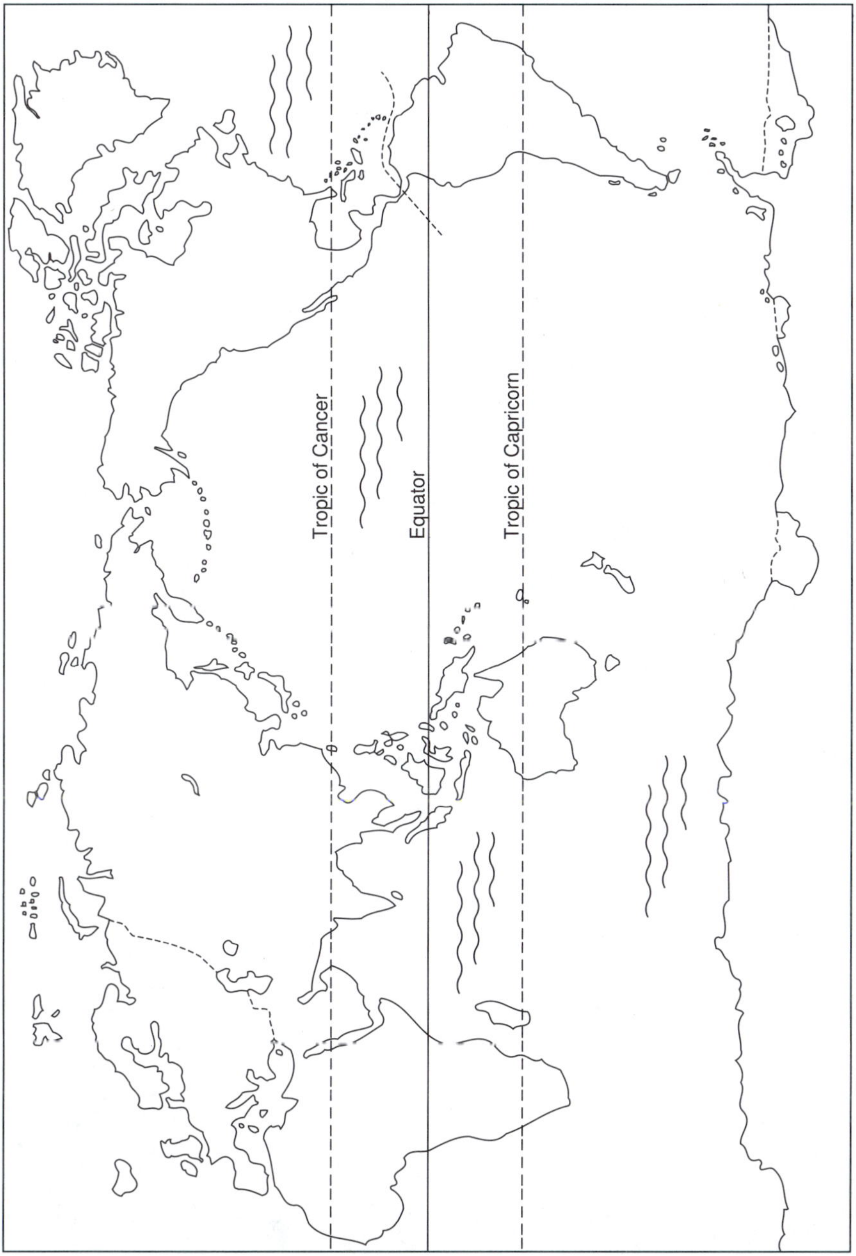

## *THE WORLD*

| Continents | Size | Population |
|---|---|---|
| Asia | 43,608,000 km$^2$ | 3,557,000,000 |
| Africa | 30,335,000 km$^2$ | 750,000,000 |
| North America | 15,285,500 km$^2$ | 454,000,000 |
| South America | 17,611,000 km$^2$ | 324,000,000 |
| Antarctica | 13,340,000 km$^2$ | (Working Scientists) |
| Europe | 10,498,000 km$^2$ | 713,000,000 |
| Australia | 7,692,030 km$^2$ | 29,000,000 |

| Oceans | |
|---|---|
| Pacific Ocean | 165,384,000 km$^2$ |
| Atlantic Ocean | 82,217,000 km$^2$ |
| Indian Ocean | 73,481,000 km$^2$ |
| Arctic Ocean | 14,056,000 km$^2$ |
| Southern Ocean | 35,000,000 km$^2$ |

| Highest Mountains | |
|---|---|
| Mt. Everest | 8,848 m |
| Mt. Aconcagua | 6,959 m |
| Mt. McKinley | 6,187 m |
| Mt. Kilimanjaro | 5,889 m |

| Longest Rivers | |
|---|---|
| Nile | 6,695 km |
| Amazon | 6,570 km |
| Mississippi-Missouri | 6,418 km |
| Chang (Yangtze) | 5,150 km |

FACTS I WOULD LIKE TO REMEMBER

# FACTS I WOULD LIKE TO REMEMBER

## FACTS I WOULD LIKE TO REMEMBER

# FACTS I WOULD LIKE TO REMEMBER